The Energy Economy

The Energy Economy

Practical Insight to Public Policy and Current Affairs

David J. Robinson

palgrave
macmillan

THE ENERGY ECONOMY

First published in 2015 by
PALGRAVE MACMILLAN®
in the United States—a division of St. Martin's Press LLC,
175 Fifth Avenue, New York, NY 10010.

Where this book is distributed in the UK, Europe and the rest of the world, this is by Palgrave Macmillan, a division of Macmillan Publishers Limited, registered in England, company number 785998, of Houndmills, Basingstoke, Hampshire RG21 6XS.

Palgrave Macmillan is the global academic imprint of the above companies and has companies and representatives throughout the world.

ISBN: 978–1–137–47169–7

Library of Congress Cataloging-in-Publication Data

Robinson, David J., 1966–
The energy economy : practical insight to public policy and current affairs / by David J. Robinson.
pages cm
Includes bibliographical references and index.
ISBN 978–1–137–47169–7 (hardback : alk. paper)
1. Energy industries—United States. 2. Energy policy—United States. 3. Economic development—Environmental aspects. I. Title.

HD9502.U52R6183 2015
333.790973—dc23 2015002992

A catalogue record of the book is available from the British Library.

Design by Newgen Knowledge Works (P) Ltd., Chennai, India.

First edition: August 2015

10 9 8 7 6 5 4 3 2 1

To my parents whose love and support made all things possible

Contents

Contents

Figures and Tables

Figures

Tables

Preface

America's economic development is a top public policy priority. A lack of federal consensus on an overall economic development strategy in Washington, DC has been filled in by regions and states across the nation. Successful local and state economic development strategies center around basic or building block strategies and more advanced five drivers strategies centering on specific industry sectors. Most communities adopt land use regulations for zoning, annexation, and eminent domain to provide sites for development. Most communities also address workforce needs of companies, use tax policies to retain or attract companies, and address larger quality of life issues to keep those companies and their workforce in town once they get them. More successful regions and states go beyond the basic building block strategies and dig deep into targeted, high-growth industry clusters such as energy, technology, global companies, advanced services, and manufacturing.

This book is a discussion about the most successful of the five drivers of economic development-energy. This book focuses on the creation of an energy led economic development strategy. It includes a basic discussion of sources of energy and overall energy policy, but the focus is on how the discoveries of sources of energy can spur massive economic growth and how the cost and availability of reliable sources of energy impact economic success as well.

As with my first book, *Economic Development from the State and Local Perspective*, *The Energy Economy* is a practical discussion of how regions and states are development successful economies around the energy sector. This book also acknowledges the multi-disciplinary approach to economic development through both the topic selection and the case study approach. No one profession owns economic development. Government workers with public administration degrees, lawyers, architects, economists, developers, bankers, and others all make up the economic development profession. Also, this book takes an "all energy is good" perspective. Renewable and nonrenewable sources of energy can help build a successful regional economic development strategy. Too often supporters of either renewable or nonrenewable sources spend too much time battling each other rather than promoting their own energy source. More importantly, if regions and states limit themselves to renewable or nonrenewable energy sources, they may be missing an energy opportunity in their backyard. As an example, shale developments are creating an economic boom across the United States, but only in a select number of regions. The same goes for wind, solar, geothermal, and biomass sources of energy that are available in different sections of the United States to varying degrees. As an example, the southwestern desert, with its warm temperatures and high number of sunny days, is a strong candidate for solar power. Going after all sources of domestic energy is the most effective way to create American energy independence as well as to drive down energy costs by creating a larger supply of energy.

Chapter 1, "Energy Led Economic Development," provides a general overview of what energy led economic development is and how it can transform a regional and state economy. The economic challenges of the state of Michigan and Nevada are used as examples. Energy Led Economic Development provides an overview of economic development, national energy policy, and energy resources supplies and the projected demand for these resources. Energy Led Economic Development outlines approaches to reduce energy costs through energy efficiency, resource demand, utility regulation, and competition and addressing urban sprawl through land use regulation. Energy Led Economic Development addresses how to build a global urban energy center, renewable energy economy, and shale oil and natural gas rural economy. Energy Led Economic Development provides an overview of what is covered in the balance of the book.

Chapter 2, "Economic Development and Energy Policy," outlines how America has approached economic development, what was and is America's approach to Energy policy, and how economic development and energy policy interact. Economic Development and Energy Policy provides a detailed history of the federal government's approach to economic development, from the days as a colony to today's state and local government dominated world. Federal energy policy is reviewed as a critical part of the review of energy led economic development strategies. How the federal government regulates energy and how they have done so over time impacts the efforts of regions and states to create energy jobs. The review of the federal energy policy illustrates how different sources of energy have been favored and then disfavored by policymakers and markets.

Chapter 3, "Energy Supply and Demand," addresses the supply and demand of all forms of domestic energy. Energy Supply and Demand defines what the sources of energy are, where they come from, and how they are used. The supply of energy is different in various regions of the nation and impacts the incentives and mandates used to retain and recruit energy companies and companies in search of cheap energy. Demand for energy includes the demand for energy on a domestic and global scale as well as demand for specific sources of energy broken down into residential, commercial, and industrial uses. With a demand for energy, promoting the development of sources of energy as an economic development tool would fail.

Chapter 4, "Energy Led Economic Development Case Study," takes the reader through a case study that will be used throughout the rest of the text. The case study involves Michigan and Nevada with a focus on urban and rural communities working on an energy led economic development strategy with renewable and nonrenewable energy sources. Energy Led Economic Development Case Study takes the reader through local and state public policy debates on a range of energy issues and develops the reader's writing, speaking, and critical thinking skills. Energy Led Economic Development addresses current topics such as urban growth boundaries, natural gas pipelines, energy efficiency, urban energy districts, energy workforce development, municipal power projects, and energy projects including wind, biomass, shale oil, natural gas, and solar.

Chapter 5, "Energy Conservation and Efficiency," addresses a specific element of energy led economic development and it discussed energy conservation and efficiency. The cost of energy is a critical element in the development of an energy led economic development strategy. One prime tool to address energy costs is to implement energy conservation and efficiency measures. Energy Conservation and Efficiency address consumer education at the federal, state, and local level, including a discussion of the Energy Star program. Energy Conservation and Efficiency also address how the government uses primarily state government energy efficiency mandates through regulation of electric and natural gas utilities and a range of financial and tax incentives encouraging energy efficiency and conservation from the local, state, and federal government.

Next, chapter 6, "Energy as Critical Infrastructure," addresses the role of energy as economic development infrastructure. Energy, its cost and reliability, impacts all economic development. High-wage manufacturers and high-tech service companies are dependent upon the availability of low-cost, high-reliability electric and natural gas service. Energy as Critical Infrastructure addresses the use of government regulation and incentives to create low-cost power. State monopoly regulation, economic development utility rate riders, and the deregulation of utility service are all covered in Energy as Critical Infrastructure. Energy as Critical Infrastructure also discusses the availability of new sources of energy in a region and state can also drive down energy costs as the market becomes flooded with a new supply of energy sources.

Chapter 7, "Land Use and Energy Led Economic Development," addresses the interaction of energy led economic development and a region's land use strategy. Land Use and Energy Led Economic Development outlines how urban sprawl drives up energy and infrastructure costs as commuters move further and further way from their workplace. The economic impact of Millennials is discussed, as well as how they can create and impact successful energy led economic development strategies by promoting mixed-use developments. Land Use and Energy Led Economic Development discusses how compact developments and urban redevelopment strategies such as Smart Growth, Brownfield redevelopment, historic preservation, Transferred Development Rights, land banks, energy zones, Transit Oriented Development, regional planning, and urban growth corridor strategies are discussed as solutions to urban sprawl.

Chapter 8, "Building an Urban Energy Center," looks to how a region can build a, global energy center. Houston is the focus of Building an Urban Energy Center and its history and current energy led economic development strategy turned Houston into a global energy success story. Houston's pro-growth land use strategies and critical infrastructure investments to build distribution connections to the world capitalized on the discovery of oil over 100 years ago. Houston's example illustrates that the production of energy can be used to build a global urban energy center through infrastructure investments that connected Houston to a distribution and logistics network, adoption of free enterprise land use policies in zoning and annexation matters, economic development oriented tax policy, and workforce policy attracting immigrant workers.

Chapter 9, "Building a Green Economy," reviews how renewable energy can build a successful energy led economic development strategy. Building a Green Economy starts with a discussion of the economic and environmental benefits of renewable energy. The long list of local, state, and federal financial and tax incentives and regulatory mandates are discussed as to how they can be used to promote the creation of a market for renewable energy market and the recruitment of renewable energy companies. Oregon is used as a model for how a state can transform its economy into an energy leader. Oregon's market for renewable energy sources and their use of mandates and incentives all built a strong economy based upon renewable energy sources.

Chapter 10, "Shale Energy Revolution," discusses the rural economic impact of oil and natural gas gained from shale rock. The economic "boom" from oil and natural gas deposits gained from the "fracking" of shale rock with horizontal drilling techniques is defined, but so is the potential economic "bust" that could create "ghost towns" after the resource is gone. Shale Energy Revolution discusses strategies "bust" preventions strategies based upon the best practices of North Dakota and other shale boom regions such as regional planning, infrastructure development through impact fees, Industrial Revenue Bonds (IRB), Special Assessments, Transportation Improvement Districts (TID) and Tax Increment Financing (TIF), economic diversification, broadband, low-cost energy strategies, and energy demand side programs.

The Energy Economy is based in the fundamental economic development practices used across the United States and illustrations of the good and the bad are discussed through a long list of best practices used across the United States. *The Energy Economy* is a practical guide that regions, states, and private industry can use to transform their economy. *The Energy Economy* is also an important learning tool for public policymakers and students to use so as to better understand how the energy sector works and how it can improve an economy.

Chapter 1

Energy Led Economic Development

Chapter Goals

1. Recognize the economic value of energy to a region, state, and nation.
2. Understand the different sources of energy.
3. Understand the regional American economic make-up of the nation.
4. Recognize the elements of an energy led economic development strategy.
5. Be able to discuss what constitutes economic development.

Michigan was home to the "American Dream." The post–World War II boom turned Detroit into a global automobile manufacturing leader, and the city of Detroit and its 1.8 million-strong population benefited. Starting in the 1970s, Michigan's American dream began to unravel. American manufacturing began to lose its market share, inflation hit, and the Arab oil embargo sent shockwaves through the American economy. Fast forward to the twenty-first century and Michigan is struggling with only 10 percent of its population in manufacturing, and Detroit lost a million people and is trying to survive bankruptcy. Michigan is searching for an economic answer, and that answer lies in energy.

Las Vegas may well be headed the direction of Michigan. Dominated by one heavily unionized industry and feeling the forces of domestic and global competition, Las Vegas took the last recession hard. Southern Nevada boomed with gaming and real estate, positioning itself as a global entertainment and retirement center. As gaming grew, other places and the overbuilt real estate market hit bottom—Las Vegas and the state of Nevada realized they had a problem. Energy may well be the answer for Las Vegas as well.

Energy is the answer for many growing regions and states across the United States. The development of new energy resources from renewable and nonrenewable sources creates an opportunity for energy led economic development strategies to build long-term wealth creation for both big and small towns. Energy led economic development strategies promote the use of locally derived energy through regulatory mandate, incentives, and demand-oriented programs to recruit high-wage jobs with companies looking for low-cost energy and energy companies building corporate headquarters, energy processing facilities, and other service and manufacturing facilities. Energy led economic development tactics involve preventing urban sprawl and promoting compact developments to address land use energy costs, promote the development of locally derived renewable and nonrenewable sources of energy, recruitment of companies in the energy industry and companies looking for low cost energy, and the building of global energy centers in major cities as well as in small towns through infrastructure development, regulatory mandate and incentives, workforce, tax, and quality of life strategies to create long term economic success. Without an energy led economic development strategy, an energy boom will be just a blip in an American economy that falls back down when the energy resource dwindles, an accident creates concern with its use, or if regulatory policy or global market conditions change.

The American Economy Broken into Pieces

Addressing American economic difficulties has been a national priority since the founding of the nation, but is more of a challenge since the 1970s. Starting as a British colony, the United States was founded to provide religious liberty for its settlers and economic benefit for the Queen. Business has always been central to the American experience. It was not until the end of World War II that the United States began to dominate the world economy. A brutal world war fought in several continents not only left millions dead, but also decimated all global economic powerhouses other than the United States.

The US factories and businesses came out of World War II unscathed, and ready to provide the goods and services the world wanted. The 1950s and 1960s were a booming economic period for the United States. While the American government was battling communism in what is termed the Cold War, American business served not just an energized American consumer, but consumers all over the world as well. America dominated global manufacturing. These manufacturing companies provided high-wage jobs requiring just a high school degree. The "Greatest Generation," which survived the Depression and fought World War II, was living the American dream on the backs of the manufacturing industry. American business leaders made generous deals with organized labor. American policy makers continued spending money on a social safety net, not just for the poor, but now also for the middle class that were looking for retirement income with social security, an interstate highway system, and funding for former soldiers to go to college on the GI Bill. Life was good.

However, as with all things, nothing lasts forever. As America entered the 1970s, the dominance of the nation's manufacturing sector began to falter. Efforts to rebuild former enemies like Germany and Japan had taken hold, and American companies started losing global economic battles. Adding to the challenge were American energy issues. The 1970s Arab oil embargo sent shock waves through the American economy and was part of a startling period of inflation that punished the American consumer and businesses alike. Global competitors for steel, automobiles, textiles, and other manufacturing businesses began to gain in US market share.

That global battle continues today. America is still the world's leading manufacturer, but China is closing in fast and should surpass the United States soon. Innovation made the American manufacturer much more efficient. It only takes 170 people now to make what it took 1,000 people to manufacture in 1950. The total number of Americans working in the manufacturing industry hovers at around 10 percent now, and millions of jobs in this industry have been lost since 1970. Thus, manufacturing still matters, but it is not the sole answer for successful regional economies.

In fact, the American economy really can be thought of not as one economy, but as a collection of seven economies with three global city-states (see figure 1.1).

These nations battle each other for jobs while they also battle global competitors. There is no interstate coordination, but American cities and

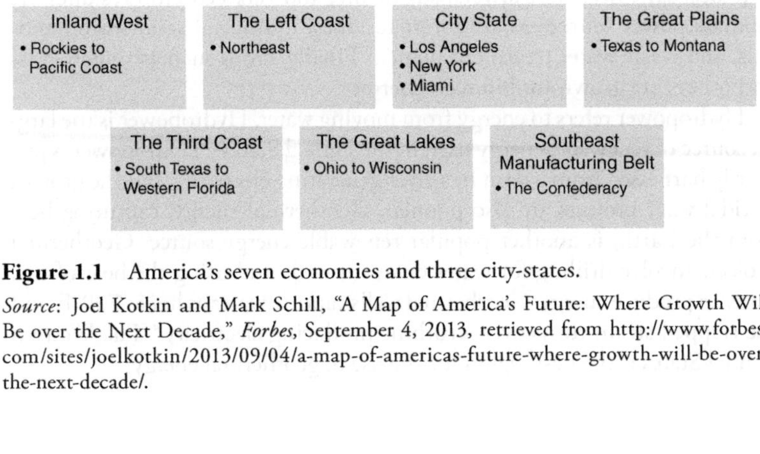

Figure 1.1 America's seven economies and three city-states.

Source: Joel Kotkin and Mark Schill, "A Map of America's Future: Where Growth Will Be over the Next Decade," *Forbes*, September 4, 2013, retrieved from http://www.forbes.com/sites/joelkotkin/2013/09/04/a-map-of-americas-future-where-growth-will-be-over-the-next-decade/.

states instead fight each other to keep the jobs they have and to steal the jobs of their neighbors. More interestingly, the winners among these seven economies and three city-states often are successful based upon the development and processing of energy. Energy, both renewable and nonrenewable, is central to the Inland West region. The Great Plains, with its Dakota shale boom, is becoming a national economic leader as never before. The Third Coast is America's traditional energy center and is booming and the Great Lakes are showing signs of life again based upon the shale boom. Regions such as the northeast are struggling, in part, because of their high cost of living, but also because they lack an energy play. The southeastern manufacturing belt is benefiting from reductions in energy costs due to the current American energy revolution.

Successful regional and state economies in all parts of the United States are built on one or more of the five drivers of economic development—the energy, technology, manufacturing, global, and advanced services industries. None of these five drivers of economic development have a wider impact than the energy industry.

Sources of Energy

America gets energy from renewable and nonrenewable sources. Renewable sources include biomass, hydro, geothermal, wind, and solar energy. Biomass is the transfer of energy from plants to power other items. Biomass products include agricultural and forest residues, municipal solid wastes, industrial wastes, and terrestrial and aquatic crops.[1] Biomass can play the role of an alternative to oil. Wheat, corn, soybeans, cotton, barley, sorghum, barley, oats, rice, rye, canola, dry edible beans and peas, peanuts, potatoes, safflower, sunflowers, sugarcane, and flaxseed crop residues are biomass power sources as are forest residues, municipal solid waste landfills, and waste water treatment plants.[2] Finally, crops such as switch grass and others are grown for biomass energy.

Hydropower refers to energy from moving water. Hydropower is the largest source of renewable energy used in the United States.[3] Hydropower is primarily harnessed from a dam in a river generating electricity, but the growth of tidal wave projects are also popular. Geothermal energy, capturing heat from the Earth, is another popular renewable energy source. Geothermal projects involve drilling for water or steam wells and using the heat of this water to produce energy. Geothermal wells capture water as hot as 700°F, and the trapped steam turns over a turbine producing electricity.[4] The Western United States is the best region for the use of geothermal energy.

Solar energy is another source of renewable energy. Solar energy uses the power of the sun as fuel. Solar technology, either passive or active, is impacted by how much sun is shining in a particular day. Solar panels capture the sun's rays through semiconductors to produce electricity.[5] Solar panels capture energy through photovoltaics, converting light to electricity, heating and cooling systems, concentrated, utility scale solar power arrays, and individual solar lighting units. All regions of the United States can capture solar energy, but the southwestern desert is the best candidate for solar power, with its high mix of sunshine and heat.

Wind captures the uneven heating of the atmosphere, irregularities of the earth's surface, and the rotation of the earth to create energy.[6] Wind is an intermittent power source and it creates energy through the use of wind turbines that converts kinetic energy into mechanical power.[7] Utility-scale wind turbines can be 100 kilowatts (kW) but may also be as large as several megawatts (MW).[8]

Nonrenewable sources of energy, which take up the largest share of energy consumption in the United States, include coal, oil, natural gas, and nuclear energy. Coal is a burnable rock formed from long-dead trees, ferns, and other plants.[9] Types of coal include anthracite, bituminous, subbituminous, and lignite. Anthracite has highest heat content and lignite the worst, but 90 percent of American coal is bituminous.[10] Coal is king in the United States as it accounts for 95 percent of the nation's fossil fuel reserves; most coal in the United States is used for electric production. Coal is America's and the world's largest source of energy, and the cheapest, but also the least suitable, from an environmental standpoint. The western United States and the Appalachian region in the east have substantial coal reserves.

Nuclear energy gained from the division of atoms creates substantial amounts of energy. Nuclear power plants collide neutrons with uranium atoms creating a chain reaction producing up to 520 degrees of energy. This heat is used to spin turbines, thus producing electricity. Nuclear energy is efficient and the United States has large reserves of the uranium needed to produce nuclear energy; however, the byproducts of producing by nuclear energy are highly harmful to humans and the potential for catastrophic injuries in case of an accident makes its use controversial.

Oil is another fossil fuel extracted from the earth and it powers the American transportation system. Crude oil is a liquid hydrocarbon mixture found underground and it serves as the basis for the production of petroleum used to create gasoline for cars and for many industrial processes. The growth in domestic shale-based oil reversed the dominance of foreign sources of oil for American use. However, oil is a carbon-heavy fuel source with substantial environmental challenges.

Finally, natural gas is a nonrenewable fossil fuel formed from buried plants and animals and obtained by drilling into the Earth where natural gas deposits have formed. Energy is produced when natural gas is burned to produce steam that powers turbines that generate electricity.[11]

America is a leader in natural gas production, and, again, recent shale-based developments have produced more natural gas than was previously thought to exist.

Shale-based oil and natural gas is produced by a hydraulic fracturing process that uses a drilling technique in which water, chemicals, and sand are pumped into a well to unlock the natural gas long trapped in shale formations by opening cracks in the rock and allowing natural gas to flow.[12] The explosion of shale oil and natural gas have dramatically improved America's domestic energy production and expanded the regions in the United States that have access to this energy source.

American Energy Usage

America's energy usage can be categorized into four major sectors of energy consumption, including industrial, transportation, residential, and commercial sectors. The Industrial sector includes facilities and equipment used for manufacturing, agriculture, mining, and construction. The transportation sector includes vehicles used for various purposes, and the residential sector includes homes, condominiums, and apartments. Finally, the commercial sector includes offices, malls, stores, schools, hospitals, hotels, warehouses, and restaurants. The American industrial sector dominates energy consumption, taking over 30 percent of the nation's energy. The transportation sector is a close second with just under 30 percent of the nation's consumption. The residential sector is third with just over 20 percent of US energy consumption followed by the commercial sector, which is last in energy consumption at just under 20 percent.

American energy consumption nearly tripled from 1949 through 2012, as the post–World War II boom dramatically increased the nation's per capita income.

America gets its energy from a range of sources, including:

- 36 percent from petroleum;
- 27 percent from natural gas;
- 18 percent from coal;
- 9 percent from renewable energy; and
- 8 percent from nuclear electric power.[13]

Nonrenewable energy sources constitute over 80 percent of all American energy usage, and, if nuclear power is included, fossil fuels are the source of over 90 percent of American power consumption.[14]

America Needs All Sources of Energy

The world runs on energy. The supply of energy comes from nonrenewable fossil fuels such as coal, natural gas, nuclear substances, and oil, but also from renewable sources such as wind, solar, hydropower, geothermal, and biomass. America needs all the sources of energy they can capture and use. That need is not based upon the overall demand for energy in the United States, but based upon the opportunity that regions and states have to retain and attract energy intensive industries and their high-wage jobs based upon a locally created energy source. That need is also based upon the global demand for energy in developing nations, where the energy demand is skyrocketing. The global energy demand will increase by one-third from 2011–2035, with emerging economies in Asia and the Middle East leading this growth.[15]

Energy debates get lost in the battle between nonrenewable fossil fuels and renewable fuels. Fossil fuels enabled the Industrial Revolution, and are still the dominant source of energy for the United States. Renewable sources are cleaner and have the benefit of not being limited. As table 1.1 illustrates,

Table 1.1 Growth potential for advanced energy sources

Wind	Increase at a rate of 6 percent to $87,000,000,000 in 2015, after being valued at an estimated $65,000,000,000 in 2010
Hydro	Estimated at $62,000,000,000 in 2010, but is expected to increase at an rate of 3.5 percent to reach nearly $74,000,000,000 in 2015
Solar	Highest rate of growth in the next 5 years with an increase of 17.1 percent, rising from an estimated $44,000,000,000 in value in 2010 to $97,000,000,000 in 2015
Biofuels	Increase in value from $45,000,000,000 in 2010 to $60,000,000,000 in 2015, for an increase of 6 percent
Geothermal	An increase of 8 percent, increasing from $6,000,000,000 in 2010 to $8,800,000,000 in 2015
Ocean	Rise from $2,400,000,000 in 2010 to $4,400,000,000 in 2015, at a healthy increase of 12 percent

Source: BCC Energy Report, retrieved from http://www.bccresearch.com/report/renewable-energy-markets-egy049b.html.

growth rates for renewable energy sources is estimated to expand. Critics point out that renewable power sources are often more expensive and do not have the energy intensity of coal and other fossil fuels. Critics of fossil fuels will point out that they create substantial environmental damage. The critics are all correct, but the nation cannot succeed without a large supply of energy. Energy is the leading source of economic success across the nation.

American Energy Policy

Reviewing the American energy policy is like watching a yo-yo. Based upon technological innovations, accidents, and regulatory changes, different energy source move up and down in their usage and popularity. However, as America became an industrialized nation, the energy yo-yo continued to move. The Industrial Revolution in the late nineteenth century was run on energy. As governments began to step in to regulate the national economy in the same period, energy did not escape the governments' review. Policymakers backed by the courts determined that electric and natural gas services were "natural monopolies." In a natural monopoly, a single firm is determined to be able to supply the market more efficiently due to the industry's average cost of production falling sharply over its output.[16] Electric and natural gas capital and service costs were determined to be so expensive that it was thought that multiple providers of the service would drive up costs. State governments stepped in to regulate electric and natural gas services as a natural monopoly service, in order to protect consumers and develop efficient electric and natural gas companies. Government regulators award electric and natural gas utility companies with a monopoly franchise for a specific geographic region and require the company to serve all the customers in that region for a reasonable price set by the regulators. The electric and natural gas company in turn gains a guaranteed rate of return for its operations.[17] Furthermore, cities began to regulate the operation of electric and natural gas utilities in public rights of way to ensure these companies could connect their services to residents and businesses, as well as to protect public land.[18] The early twentieth century saw a new development when municipalities themselves got into the electric and natural gas business as the Progressive Era took hold and the popularity of large, private companies waned.[19]

Beyond government regulation, energy market transformations occurred based upon the discovery of new sources of power. The twentieth century American industrial boom mandated large energy production from more efficient sources of power. Wood and other biomass sources gave way to coal. The plethora of oil discoveries was timed perfectly with the invention of mass production of the automobile, that led to lower prices for cars and massive expansion of the automobile sector. Again, policymakers did not sit idly by as the energy industry grew. The federal government in the early 1900s began to regulate the oil industry's interstate pipelines; however, the early oil industry, unlike electricity and natural gas, benefited from a lack of direct regulation by the government.[20] That all changed, however, when the government filed an anti-trust lawsuit against Standard Oil and the court broke up one of America's largest companies into a series of smaller, regional oil companies. New Deal Era legislation regulated the size and geographic service area of electric and natural gas utilities and required them to register with the Securities and Exchange Commission, and interstate electricity transmission was regulated as well.[21] Also, rural electric and natural gas service was addressed with federal legislation, which provided low interest government loans for utility companies to serve these expensive markets.[22] Following World War II, America entered the atomic age and federal policymakers again regulated this energy source to ensure that the development of nuclear capabilities and power management would be under civilian, not military, control and would remain in the United States.

Consumer protection is not the only basis for the American energy policy. The American energy policy is also based upon a strategy of energy diversification. The federal government and its state government partners worked to license a range of water, oil, and natural gas companies to support their growth and development. However, the energy diversification strategy failed to address the complete reliance of auto industries on oil to power cars. Post–World War II America was a time of unprecedented wealth for the only industrialized nation left to manufacture products for the rest of the world. The miles driven by Americans grew as along with the thickness of their wallets, as a suburban sprawl took hold and put Americans in their large, gas burning cars for increasingly longer drives. All was fine until the 1970s, when oil consumption reached well beyond what the United States could produce and foreign powers in the Middle East were providing a large share of oil that kept the US economy working. Foreign policy conflicts then sent the American economy into a shockwave when the Arab nations in the Middle East stopped selling oil to the United States. Gasoline lines and skyrocketing prices stopped the American economy. The federal government responded to this energy challenge. A series of federal laws promoted greater automobile efficiency, taxed gas-guzzling cars, promoted alternative energy development, and created a Strategic Petroleum Reserve to stockpile oil in case of a national emergency.[23]

Congress also aggressively worked to promote the use of coal to substitute for the use oil whenever possible.[24]

As the 1970s ended and the Reagan Revolution of the 1980s took hold, America moved again, in a different direction, with its energy policy. Conservative policymakers saw the answer for an energy policy in the marketplace and moved to begin the deregulation of the electric and natural gas industry. Federal law set up the framework for wholesale electricity generation deregulation, and then state regulators began to deregulate retail electric and natural gas services. The age of energy deregulation took hold, and now nearly half of all the states in the nation have competitive electric and natural gas retail services. Energy diversification efforts continued as America entered the twenty-first century. Conversations that started in the 1970s around alternative energy sources took hold some forty years later, as renewable energy sources were favored by state and federal policy makers through a series of mandates and incentives to increase renewable energy production use and production.

Energy Led Economic Development

Economic development involves the use of public subsidies to encourage private sector wealth creation.[25] Local and state government and private sector leaders work collectively to promote the creation of high-wage jobs and capital investment in their communities through a range of tactics centered primarily around business retention and attraction campaigns. Regional and state economic development strategies center on basic or Building Block or advanced Five Drivers of Economic Development strategies. Building Block economic development strategies include land use planning, annexation and eminent domain to prepare sites, infrastructure being developed, workforce trained, tax policy addressed, and quality of life issues being addressed. Most communities conduct economic development operations with these Building Block strategies. However, regions and states leading the nation in its economic recovery do not stop with Building Block strategies, but extend into more aggressive, pro-active Five Drivers of economic development strategies. These Five Driver strategies center on core industry growth in energy, technology, advanced service and manufacturing, and global firms. These industry clusters have a strong track record of creating high-wage, private sector jobs. Of all these industry clusters, energy leads the nation in the number of regions and states it is affecting positively from a business growth standpoint.

Energy led economic development strategies recruit energy companies and companies seeking low-cost energy through the use of regulatory mandates, incentives, and demand oriented programs to recruit high-wage jobs. These companies could be energy processing, corporate headquarters, heavy industry, or high-tech companies dependent upon reliable and cheap energy. Energy is connected to economic development as a critical infrastructure source as well as a major industry—providing advanced services, manufacturing, and research and development jobs. Energy led economic development is the strategic approach to turn energy into more than just a natural resource. Energy led economic development is an approach to build a regional and state economy where energy and its related industries are high-wage job creators. As figure 1.2 illustrates, energy led economic development strategies include several stages.

Regions and states have to make challenging decisions related to the regulation of energy extraction operations. Whether the energy extraction project is of shale oil through a fracking process or placement of an industrial windmill, proponents and opponents to the placement of these facilities always exist, And the same goes for the processing facilities. Opposition can be ideologically based or just a neighbor unhappy with a land use change occurring next door. Local and state policy makers adopt regulations to ensure the energy extraction process is undertaken safely. Failure to address these safety concerns can result in the death of a booming energy source. Who can forget how the Three Mile Island accident in the 1970s stopped the growth of the American nuclear power industry!

Regions with a new source of energy being extracted from the land have an opportunity to utilize not just a new source of energy, but this energy explosion should also drive down the cost of energy in that community. In essence, the flowing of natural gas and oil from shale development drove down the overall cost and regional cost of this fossil fuel. In addition, the

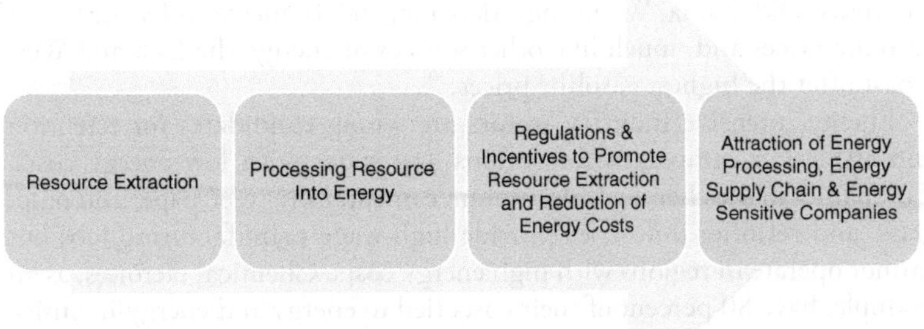

Figure 1.2 Stages of energy led economic development.

extraction of the energy source gives that region an opportunity to give birth to the energy industry.

Low Cost Power as an Economic Development Driver

Americans use 19 percent of the world's energy and this energy usage is driven by the industrial nature of the American economy.[26] From an economic development standpoint, not all states have equal levels of consumption and costs for energy. The large, industrialized states of Texas, California, Ohio, New York, Pennsylvania, Illinois, North Carolina, Michigan, and Indiana top the US states in energy consumption. Cold weather northern states use more natural gas to heat their homes and businesses while southern states consume substantial amounts of electricity to cool these same types of facilities in the warm summers.

Beyond levels of consumption, the cost of energy is a top factor influencing the corporate site location decisions of companies considering where to locate their facilities. Energy availability and costs rank in the top ten of corporate site location consultant factors. The cost of energy varies among the different states and it varies with sources of energy as well. Idaho, Wyoming, Washington, Utah, and Kentucky offer the best electric rates, and New Hampshire, New York, Alaska, Connecticut and Hawaii offer the highest.[27] The cost and reliability of a region's transportation fuel impacts on transportation and logistics firms, as America consumes 27 percent of its energy in transportation.[28] Oil constitutes 93 percent of America's transportation costs and small business owners found high fuel costs to be an impediment to economic growth.[29] South Carolina, Arkansas, Oklahoma, Wyoming, Alabama, and Tennessee offer the lowest gasoline prices and, much like other sources of energy, the East and West Coast offer the highest gasoline prices.[30]

Energy intensive industry sectors are strong candidates for retention and attraction campaigns for regions and states with low energy costs. Companies in the chemical, aluminum, cement, iron, steel, paper and pulp, glass, and refining industries provide high-wage manufacturing jobs but cannot operate in regions with high energy costs. Chemical factories, as an example, have 80 percent of their costs tied to energy and energy-intensive industries constitute just over two-thirds of all global energy use.[31]

Both nature and policymakers impact the energy cost of a region or state. Most regions in America have access to an affordable source of energy based upon available natural resources such as water, geothermal, shale, oil, natural gas, wind, solar, biomass, oil, or natural gas. The Tennessee Valley Authority (TVA) is an example of a multi-state region who capitalized on federal resources to capture the power of water to provide low-cost energy. Today, the TVA serves 155 government cooperative power distributors and directly serves 57 large industrial or government installations.[32] The TVA is an economic development giant that plays a major role in retaining and attracting high-energy companies with power costs 16 percent below the national average stimulating $5,900,000,000 in business investments. Shale oil and renewable sources have a chance to play the same role as the TVA for the regions having access to these power sources.

Whether a region has direct access to a large pool of energy from their natural resources or not, state policymakers can decide whether their regions have low or high costs for power. Traditionally, the costs of power are driven by regional policy decisions regarding whether to adopt a regulatory or deregulatory model for electric and natural gas services. Either approach can produce low-cost power. Utility regulatory models that treat electric and natural gas companies as a monopoly permit these companies to offer services without competition, but at a previously agreed upon price. Even those states that regulate utilities as a natural monopoly offer special economic development incentives for energy-intensive companies. These companies gain reduced rates for electric and natural gas services in exchange for the promise of job creation and capital investment. The larger the job creation and capital investment—the larger the energy price reduction for energy-intensive companies. Nearly half the states have deregulated power utility services and have left it to the market to set retail rates for electric and natural gas services. A great debate is raging between policymakers about whether the regulated or deregulated approach creates lower energy costs. Either approach can work but policymakers have to make challenging choices when it comes to driving down energy costs.

Energy Efficiency Addresses High-Energy Costs in High-Wage Industries

Promoting energy efficiency is another method to reduce energy costs. Companies using less energy drive down the overall demand as well as reduce their own energy costs. Three energy efficiency strategies center on consumer education, use of economic development incentives, and development of product standards. Energy efficiency education campaigns empower consumers to drive down their energy costs. The federal government created the Energy Star designation for products lets consumers know how a particular product is ranked from an energy consumption standpoint. Energy Star product and service ratings are certified by US Environmental Protection Agency (US EPA) through a third party certification program.[33] The Energy Star program creates substantial energy efficiency-based savings. Over $200,000,000,000 in energy efficiency savings have been gained, based upon the use of Energy Star products and services.[34] Energy audits are another energy efficiency program based upon educating consumers. An energy audit identifies where and how much energy is consumed and how energy efficiency steps can reduce costs.[35] Energy audits review energy bills to determine if a facility's costs are out of line with energy efficient facilities.[36] Windows, water heaters, boilers, HVAC, and other major power users are reviewed for their power usage.[37]

Thousands of local, state, and federal tax incentives exist to promote energy efficiency investments. Tax credits, tax abatements, grants, and loans are offered across the nation to subsidize energy efficiency investments. Technology exists in many industries that can help a company spend less on energy, but the cost of the technology may exceed the payback in utility services. Tax incentives can make an energy efficiency investment pay dividends. Examples of energy efficiency based tax incentives include: corporate tax incentives, grant programs, green building incentives, green company incentives, green equipment and facility public finance tools, performance-based incentives, personal tax incentives, property tax incentives, rebate programs, and sales tax incentives. Finally, energy efficiency programs can include regulatory mandates to require the use of alternative energy sources. If tax incentives are a policy "carrot," regulatory mandates such as requiring that a utility provide rebates to manufacturers making energy efficiency investments are the "stick." Regulatory mandates can also be utility "demand side" programs to require utility companies to reduce energy costs by a certain percentage by working with energy intensive companies to reduce their power consumption. Utilities can also be required to implement energy efficiency Integrated Resource Plans, where utilities receives a shareholder incentive, recovery of lost revenue, or other methods of cost recovery for additional expenses related to the implementation of energy efficiency programs.[38] Minimum efficiency standards for appliances are an example of an energy-efficiency mandate.[39] State building codes are another energy efficiency mandate used to enforce green building standards to require energy efficient construction approaches and the use of renewable energy sources.[40] Governments also can lead by example by requiring all government structure to be certified as energy efficient based upon national standards.[41] Governments can also mandate their own use of energy efficient equipment by requiring governments to use only energy efficient rated equipment.[42]

Land Use Regulation Impacts Energy Led Economic Development Strategies

America's land use policies impact energy cost and consumption. America's post–World War II urban sprawl development pattern contributes substantially to rising energy consumption and costs. Zoning and interstate highway developments spurred America's cities to grow out, not up. Promoting more compact suburban development and redeveloping America's urban centers can drive down energy costs and help build an energy led economic development strategy.

Demographic changes create an opportunity to build energy led economic development strategies around land use. The American Millennial Generation, 18–36 year olds, constitutes one-third of the American population and is the nation's largest demographic group.[43] This group has substantial economic power and will constitute three-fourths of the nation's workforce by 2025.[44] Millennials are different consumers from older generations. Millennials own fewer cars and they are much more likely to use public transit.[45] They also want to live in mixed-use developments in cities and do not aspire to own a home in the suburbs.[46] Millennials are leading an urban redevelopment revolution, as multi-family apartment units and condos pop up in cities all across the United States.[47] Finally, Millennials are supportive of the use of renewable energy sources and the use of a sustainable development model.[48] Regions and state are interested in recruiting and retaining Millennials because of their value to a workforce struggling to replace retiring Baby Boomers.[49] Energy and all companies need millennial workers. In addition, the Millennials' interest in urban, mixed use development creates an opportunity for an energy led land use redevelopment model.

A land use, energy led economic development strategy involves developing unimproved land in smarter, more densely patterned developments and adopting urban redevelopment strategies to redevelop existing sites. These tactics together will turn fewer cornfields into new homes and strip commercial retail centers. These tactics together will also consume less energy and, in some cases, utilize existing traditional and renewable energy sources already in place. More compact new developments and redeveloping urban centers are not only environmentally sensitive, but reduce regional infrastructure costs by stemming the tide of urban sprawl. The post–World War II suburban explosion led to a substantial increase in the demand for oil as the number of car miles grew to connect workers with their suburban homes and Central Business District jobs. More compact new developments and redeveloping the urban core reduces infrastructure spending and promotes energy consumption by concentrating customers around existing energy infrastructure.

Compact new developments and urban redevelopment projects are gaining favor, not just with policymakers, but also with the marketplace as a whole. Millennials are more interested in compact, mixed use developments located in cities near transit stops and not in the suburbs. Multifamily housing is booming and car ownership among Millennials is dropping. Strategies to promote urban redevelopment centers include creating land banks to redevelop abandoned housing; redeveloping Brownfields and historic sites; building energy district; and adopting Transit Oriented Development land use strategies to promote inward growth. Strategies to promote more compact development patterns will use Transferred Development Rights and smart growth regional land use strategies to manage growth by providing a boundary for growth. These tactics are creating jobs by driving down energy costs and developing properties in popular neighborhoods and regions.

Energy Industry as an Economic Driver

The energy industry is the basis for successful urban and rural communities. This is true for both renewable and nonrenewable energy sources. Energy is truly one of the few industries that not only serves a vital infrastructure supporting the development of all companies but also can serve as the economic core of rural and urban centers.

Houston, Texas is a model for the development of an urban global energy center that started with the discovery of oil 100 years ago. Houston turned into the fourth largest city in the United States based upon a booming oil industry, and the city is the world's fortieth largest economy.[50] Houston's growth is rooted in not just a discovery of oil nearby but in the business community's wise choice to gain federal funding to build a ship canal, creating a logistics and distribution operation for the global oil industry to link Houston's oil developments to the Gulf of Mexico and the world. Houston furthermore utilized their oil-based wealth to grow into a major health care center. Houston also benefited from the pro-growth annexation policies in the state of Texas that permitted them to grow. Finally, Houston's pro-growth policies extend into its land use model of "no zoning."

Fossil fuels are not the only source of energy that can spawn major urban centers on the backs of the energy industry. Renewable energy sources are building successful energy led economic development initiatives across the United States. Renewable energy sources such as wind, biomass, solar, hydro, geothermal, and other renewable sources of energy are widely available, and many regions and states have an opportunity to turn these resource sources into the locations of services, research and development centers, and manufacturing facilities in these industries. US demand for renewable energy equipment is roughly the size of America's construction machinery, farm machinery, and household appliance manufacturing industries.[51]

Oregon offers a national model for a renewable energy center, located in Portland. Oregon's strategy started with a plan spurred by the 1970s energy crisis, and the state's renewable energy strategy created 43,000 renewable energy jobs with nearly 5,000 employers, constituting 3 percent of the state's workforce.[52] Oregon grew this renewable energy industry through an aggressive array of tax incentives and renewable energy mandates. Oregon grew companies in this industry with both policy "carrots" and "sticks," and its urban growth boundary is a national, smart growth model to build sustainable land use policies. Oregon's initiative not only illustrates a successful renewable energy strategy, but also proves it can be done on the statewide level.

Finally, rural communities are the site of amazing economic growth based upon recent energy discoveries primarily in shale oil. Rural communities, with the job loss in farming, have lower incomes, higher poverty, less educated populations, and higher mortality rates than urban counterparts do. Rural communities across the nation are the centers of most shale developments, and the growth is impressive. Rural North Dakota is the site of massive shale oil exploration and its economic impact is substantial. North Dakota is the second largest oil producing state in the nation—up from ninth—and the state has the fastest growing economy in the nation and the lowest unemployment rate in the nation, all on the backs of shale oil. Small communities in North Dakota and elsewhere are working to make community infrastructure investments, complete land use planning to manage growth, build long-term housing, and capitalize on new government revenue to enhance the region's quality of life through the development of recreation centers and other community assets. Small communities are booming because of the oil gushers, but are also focused on managing the "oil bust" that will no doubt follow.

Will Las Vegas Be the Detroit of the Next Generation?

Defining a region's economic challenge is the first step in developing an energy led economic development strategy. The challenges of Detroit are well known, but the once "boomtown" of Las Vegas is not widely discussed.

A larger question is *will the next generation grow up knowing the booming Las Vegas or will Las Vegas be seen as an economic loser—over built with housing and a massive over concentration of one industry-gaming?*

The Rise and Fall of Detroit

In 1701, French military leader Antoine de la Mothe Cadillac traveled in 25 canoes with 50 soldiers from Montreal to set up a military and trading outpost now known as Detroit.[53] Before Detroit was a global automotive manufacturing center, the region had a strong start on buggy manufacturing.[54] Detroit built a buggy cluster of companies that led entrepreneurs such as Henry Ford, the Dodge Brothers, Walter Chrysler, and others to launch the automobile revolution in the early 1900s in Detroit. This system of mass production turned Detroit into a truly global powerhouse. All this industrial success led to population gains for Detroit. The city itself started with 285,000 people in 1900 and by 1950 was at 1,849,568. Detroit was competing to be America's "second city."[55] In fact, the 1950s saw Detroit with the highest median income and rate of home ownership of any major city in the nation.[56] Detroit was the home of the American dream. In 1955, GM's President Charlie Wilson is reported to have uttered the famous sentence—"what is good for General Motors is good for America." Whether he said it or not, the statement was true at the time. In the 1950s, four out of every five cars made in the world were made in the United States, and half of those were produced by GM.[57] However, starting in the 1960s, American auto companies' expensive labor agreements and lack of innovation began to catch up with them, as global competitors began gaining a foothold in the US market. When the Arab oil embargo forced Americans to look at the gas consumption of their cars, the smaller, global imports became instantly attractive. The 1970s and 1980s saw the rise of the foreign automakers. Part of what helped this initiative was the investment foreign automakers made in the United States. Honda, Toyota, and many others opened up auto assembly plants in the United States. This foreign investment directly addressed the Buy American campaign by the Big Three and helped market their cars to the American consumer.

The City of Detroit started to struggle at the same time the Big Three were starting to show signs of weakness. A party for two returning Vietnam veterans and the police response to the party led to 5 days of rioting, 43 deaths, over 1,000 injured, and over 7,000 arrests. To add insult to injury, all of America watched on television as Detroit burned before their eyes, with even the National Guard struggling to regain control of the city.[58]

A leading American city that produced unprecedented wealth as well as the strongest middle class in the country was on fire. What caused this explosion?

The final piece in the collapse of the city of Detroit was all the public policies that were framed and legal efforts that were taken to save the cities. First, the urban renewal efforts of the 1950s are a classic example of well-intentioned public policy gone bad. Local, state, and federal government leaders lead an effort to "save" cities by destroying city block after city block and replacing these classic structures with public housing and other projects. Besides destroying some great American architecture, the result was a massive concentration of poverty in urban neighborhoods that could not afford additional challenges. In addition, the Federal-Aid Highway Act cut into pieces nearly every major American city to build a national highway network that not only connected major US cities, but also instantly created a suburban market that urban centers were not ready to compete with. Finally, efforts to provide a high quality of education to African Americans through the abolition of racially segregated schools led to the courts ordering the busing of students in urban centers all across the United States as the remedy. When the US Supreme Court put a stop to this remedy at the suburban city limit, the white flight plan from many American cities was complete. Many major American cities, including Detroit, became heavily concentrated with African American and high poverty populations.

Fair or not, no one is going to rank Detroit high when it comes to promoting innovation and entrepreneurship. Many regions and states have moved in the direction of high tech, knowledge-based jobs, and company creation to build economic success. Microsoft, Apple, Intel, HP, and others are the GM, Ford, and Chrysler of the twenty-first century. The development of these companies is often accomplished with the creation of an innovation-based economic model that includes a strong link between university researchers and the private sector, available workers for specialized industries, and access to capital.

Michigan, at best, ranks in the middle of the pack when it comes to the development of a science and technology-based economy, with groups such as the Milken Institute State Science and Technology Index.[59] The Milken Institute looks to key areas in the innovation-based economy as indicators of likely success in the twenty-first century, such as human capital investment, research and development inputs, risk capital and entrepreneurial infrastructure, technology and science work force, and technology concentration and dynamism.[60]

Michigan does no better with the Small Business Foundation of Michigan, which gave the state only one star out of five for its entrepreneurial environment.[61] Michigan ranked poorly in growth in the number of small businesses, small business payroll growth, and the increase in the number of high performance firms, but better in net establishment increase.[62] From an entrepreneurial standpoint, the state ranked poorly in growing gross state product, foreign direct investment, and large business payroll, but ranked higher in private lending to small businesses, export-related jobs, patents, and private industry research.[63]

Detroit is struggling to embrace globalism. This certainly does not mean that GM, Ford, and Chrysler are not selling cars overseas. In fact, GM's operation in China is a model for many other Fortune 100 companies, and the state overall ranks fourteenth in the nation for export-related jobs.[64] However, these global investments might be better for GM's stock price than Detroit's workforce. Detroit and Michigan are prototypes for twenty-first century global economic losers. The state of Michigan, a top 10 economic powerhouse, ranks fiftieth in foreign direct investment—that means global businesses and resources are not investing in companies in the state of Michigan.[65] The state is also dead last when it comes to export growth.[66] Metro Detroit grew from 100,000 people in 1900 to 5,200,000 by 2010.[67] However, growth during the last half of the twentieth century occurred not in the urban center, where it all started, but in the outlying suburban communities. As an example, while Metro Detroit actually grew from 4,800,000 to 5,200,000 from 2000 to 2010 the City of Detroit lost 25 percent of its population.[68] In fact, the City of Detroit only makes up 18 percent of the population of Metro Detroit, which in 1960 was at 44 percent.[69]

The Rise and Fall of Las Vegas

Las Vegas began as an oasis in the desert. Two springs served as a starting point for the world's global gaming center of today.[70] Las Vegas the city began in the early 1900s—about the same time Detroit was watching Henry Ford start to build cars. Passed over by the Mormons, Las Vegas and Nevada chose another route. In 1911, the state passed a "quickie divorce" law.[71] In 1931, the state of Nevada legalized what was going on anyway and permitted gambling, and, in 1935, the construction of the Hoover Dam provided a larger source of water and energy to permit growth in the region.[72] However, Las Vegas is still a town of just over 8,000 residents as World War II began.[73]

Benjamin "Bugsy" Siegel and Willis Haviland Carrier should be regarded as the modern fathers' of Las Vegas. Post–World War II Las Vegas had gaming laws, sunshine, land, and electricity in place to spur development. It was a relatively short drive from the population centers of Southern California, and was a perfect distance for quick getaways. However, with the sunshine came several months of high temperatures. 100-degree days are not good for the operation of most businesses, that is, until the invention of Cornell University graduate Willis Haviland Carrier came to benefit regions such as Las Vegas. Carrier is credited with the invention of today's modern air conditioner.[74] Without air conditioning, regions such as Las Vegas would remain a small desert oasis.

By 1980, Las Vegas' population reached over 160,000 people.[75] However, the boom in Las Vegas had not even really started. The big boom for Las Vegas centered on an explosion in the tourism trade and gaming matched by a commercial and residential real estate explosion.

The simple fact is gaming is everywhere, but Las Vegas created the largest cluster of gaming centers in the nation. The Las Vegas gaming industry is nearly a $ 6,000,000,000 industry, and almost doubles the size of the Atlantic City market in revenues.[76] As table 1.2 shows, the top ten employers in Las Vegas are dominated by government and gaming as well.

Marching along with casino gaming, Las Vegas during the 1990s and first decade of the twenty-first century enjoyed an unprecedented commercial and housing boom. The Las Vegas office market grew from 28,400,000 square feet in 2000 to 48,500,000 square feet in 2008.[77] The Las Vegas industrial market grew substantially as well during the same timeframe—going from 66,900,000 square feet in 2000 to 102,600,000 square feet in 2008.[78] Finally, the Las Vegas home building industry boomed in the early twenty-first century with nearly 50,000 homes sold in 2000, and that number nearly doubled by 2005.[79] The City of Las Vegas is promoted as a quality retirement community on par with Florida and Arizona.

Table 1.2 Las Vegas top ten employers

1.	MGM Grand
2.	Caesars Palace
3.	Station Casinos
4.	Wynn Resorts
5.	Boyd Gaming
6.	Las Vegas Sands
7.	Walmart Corporation
8.	Cosmopolitan of Las Vegas
9.	The Valley Health System
10.	Supervalu

Source: http://www.vegasinc.com/business/public-record/2012/mar/19/list-largest-employers/.

The city is also highly unionized. While most states and regions have only a single digit percentage of their residents in a union, the gaming and convention industry in Las Vegas is actually heavily unionized. According to the Bureau of Labor Statistics, in 2010, the number of workers belonging to a union was 151,000 in Nevada, accounting for 15 percent of wage and salary workers in Nevada, compared 5.7 percent nationally as reported in 2009.[80] At its peak in 1996, Nevada's union membership rate was 20.4 percent. Nevada goes against a national trend against unionization.[81]

Las Vegas is in the same recession everyone else in America is living. As a boom town, the bust is unusually big. The question is *does Las Vegas really ever recover in the short or long term?* Las Vegas positioned itself to be the Detroit of the twenty-first century. Much like the 1967 riots precipitated the collapse of Detroit, the real estate implosion could be the inflection point that marks a quick slide down the global economic roller coaster for Las Vegas. To say Las Vegas has been hit by the "housing bubble" is an understatement.

The causes of the housing bust are many, but the reality for Las Vegas is that it created a major hole in the continued growth of the regional economy. Since 2007, Las Vegas won the award no one wants—the number one ranking for home foreclosures.[82] Following the 2001–2002 recession, the home occupancy rate was 92.8 percent and that rose to a high of 95.7 percent in 2006, but now Las Vegas has dropped down to a decade low of 90.8 percent.[83] The Las Vegas office market is not doing much better. The vacancy rate that was 7.7 percent in 2000 now stands at 23.4 percent, with 11,600,000 square feet of vacant space, and the Las Vegas industrial market that was at its peak development of 102,600,000 square feet in 2008 has only added 2,000,000 square feet since then.[84] The Las Vegas retail market is in the tank as well. From a peak of 48,500,000 square feet of retail space developed in 2007, growth slowed since, with a supply of 51,800,000 square feet today.[85] The once low Las Vegas retail vacancy rate of 4 percent in 2007 is now at 10.5 percent today.[86] Due to office, industrial, and retail space growth of 50–70 percent over the last decade, Las Vegas may need four years to work through its existing excess office inventory and three years for the retail and industrial sector.[87]

The greatest challenge Las Vegas faces is that its one and totally dominant industry, gaming, is entering a new era of competition. Gaming is good business and it is everywhere. In 2010, gross gaming revenues for commercial casinos were $34,600,000,000 and $7,590,000,000 was paid in local and state governments taxes.[88] The gaming industry also is a big job producer. 340,564 people are directly employed by the gaming industry across the United States, and they earned $13,300,000,000 in wages, benefits, and tips during 2010.[89]

Nearly 40 of the 50 states have some form of casino or video lottery terminal facilities. Twenty states have full-fledged casinos in operation.[90] Las Vegas is still the largest gaming center in the United States, but the chart below illustrates the strengths of other markets all across the nation (table 1.3).

Gaming competition for Las Vegas is coming from information age technology as well as global forces. The Internet is a new force in the gaming market. While of questionable legal status in many place, Goldman Sachs estimated that the global on-line gaming market is around $12,000,000,000 and is a likely legalization target just for the government tax revenue alone.

In addition, global competitors are not letting the United States dominate the gaming market. Most famous is the "Las Vegas of the East," also known as Macau. Like many US markets, Las Vegas is facing a stiff challenge for global tourism from China. In fact, with annual gambling revenues of more than $13,000,000,000, Macau is the largest casino gambling jurisdiction in the world.[91] Macau grew from 11 casinos in 2002 to 33 today, and many of Las Vegas' casino operators have major casinos' in this Chinese city.

Las Vegas, and indeed Nevada, are also behind when it comes to preparing its work force for employment in the information age. The heavy flow of gaming revenue the state gains could have been a strong start to build a public school system the world would admire. Unfortunately, Nevada appears to have moved in the opposite direction. While there are many good quality schools in Nevada, the state in 2009 received the distinction of having the

Table 1.3 US state gaming market

Casino Market	2014 Annual Revenues (in $)
Nevada	10,860,000,000
Pennsylvania	3,160,000,000
New Jersey	3,050,000,000
Indiana	2,610,000,000
Louisiana	2,400,000,000
Mississippi	2,250,000,000
New York	1,800,000,000
Missouri	1,770,000,000
Illinois	1,640,000,000
Iowa	1,470,000,000

Source: http://www.usatoday.com/story/money/business/2014/04/27/top-gambling-states/8168681/.

lowest high school graduate rate in the nation, according to a report from Education Week.[92] Nevada only graduated 47.3 percent of its students on time and Clark County School District's graduation rate was at 46.8 percent, placing it at 43 among the 50 largest school districts in the nation.[93] More troubling is the news that the graduation rate in Nevada declined by 23 percentage points between 1996 and 2006 while the national graduation rate improved by 2 percentage points to 69 percent.[94] Nevada is ranked forty-ninth in the nation when it comes to the number of its residents with bachelor's or associate college degrees.[95] Las Vegas is the least well-educated large metro area in the nation. The economic impact of this designation is troubling. The four-year college attainment rate of the best educated metropolitan area, Washington, DC, at 46.1 percent, is more than double that of the least well-educated metropolitan area, Las Vegas, at 20.2 percent.[96]

How Can Las Vegas and Detroit Survive?

How does Las Vegas prevent itself from becoming the Detroit of the twenty-first century? The facts illustrate that Las Vegas is a growing metropolitan center based upon one, dominant industry with a unionized workforce in a region with underperforming schools, unsafe streets, and an economy that is not geared toward the high wage jobs of the future. How can Detroit and Michigan recover? Detroit benefits from a massive infusion of federal funding, a streamlining of city government by the bankruptcy process, and a strong core of manufacturers sitting just outside the city limits. However, both regions' prime issue is their overreliance on one industry. Both Las Vegas and Detroit, and the states that host them, need to look to energy led economic development to diversify their economies and attract high wage jobs back to these states.

Both Nevada and Michigan have access to sources of renewable energy. Nevada has the sun and Michigan has a strong start on the wind industry. These sources of renewable energy can be harnessed and used to provide a new source of power, but can also be used as a tool to recruit renewable energy companies. Unfortunately, neither state is sitting on large deposits of oil or natural gas from shale deposits; thus, state leaders need to use energy efficiency and regulatory or market measures to reduce power costs for manufacturers and technology companies that the state is seeking to recruit. Urban sprawl is a major challenge that needs to be addressed in greater Detroit and Las Vegas, but these cities have the potential to be global energy centers. Small communities also can, with the development of renewable energy projects, become global energy success stories in their own right.

Value of Strategic Energy Led Economic Development Plans

Good energy led economic development strategies involve the creation of a strategic plan. Strategic planning first identifies the economic opportunities of a region by using an economic cluster analysis and producing a Strengths, Weaknesses, Opportunities, and Threats (SWOT) analysis. Next, the heart of the energy led economic development strategic plan is laid out through a discussion of the plan's goals, strategies, objectives, and tactics.

Economic cluster analysis identifies a region's business strengths and matches those strengths with the industries that are growing and successful. An economic cluster is a geographic concentration of firms in the same industry and the supporting companies and organizations that help those firms succeed.[97] Economic cluster analysis uses census and business data from reporting services or direct surveys to define the industry located in a defined location.[98] Successful economic development cluster initiatives start with a clearly defined geographic boundary and an understanding of the competitive advantage of clusters, but also focus on addressing challenges to the growth of the cluster in that region.[99] Next, a supply chain analysis provides an expanded industry perspective to identify other partners needed by those companies. A SWOT Analysis then matches up a region's industry strengths with potentially growing markets. A SWOT analysis defines who a community or state is as well as guides them on what they want to be.

Energy led economic development strategic plans include a discussion of goals, strategies needed to achieve those goals, objectives created to measure the success of the project, and tactics that detail the specific steps needed to achieve success. An energy led economic development strategic goal is the region or state's vision for what they want to become. This goal is impacted by the actual condition of the regional economy and what energy resources are available to the region. The economic development cluster analysis and SWOT analysis impact the goal production. The actual goal statement is short, direct, and concise.

Following the development of the goal statement, energy led economic development strategies address the objectives of the plan. These objectives use metrics to define the success or failure of the energy led economic development strategic plan. Measures of economic success are often used, such as household family income, unemployment rate, and poverty rates. Another metric used is shift-share analysis. Shift-share analysis compares local economic growth with national economic growth in the same industry sector. Shift-share analysis analyzes a region's growth into segments by comparing the national economic growth rate with the growth or decline of employment in that industry at the local level.[100] Shift-share studies measure how regions impact the growth or decline of employment in a specific industry and attempt to distinguish the growth of that industry sector attributable to national economic factors.[101]

Once an energy led economic development goal and clear objectives are set, the next step in creating an energy led economic development strategic plan is adopting strategies needed to meet the goal. These strategies embrace the Building Blocks and Five Drivers of Economic Development to create strategies to incentivize the creation of high-wage jobs through incentives and mandates. The final piece of creating an energy led economic development strategic plan is the creation of tactics. These tactics are specific action steps laid out in specific timeframes. This outline of action steps provides a guide to achieve the overall goals of the plan.

Big Questions

1. Will energy led economic development continue to produce major benefits?
2. What is the energy industry booming?
3. Can we grow with both renewable and nonrenewable energy sources?
4. What are the key elements of an energy led economic development strategy?
5. Is the shale boom temporary? Will it prove more harmful than helpful as time goes on?

Class Debate

The class will go through an exercise to support the creation of a strategic energy led economic development plan for Michigan or Nevada. Class discussion will focus on how energy led economic development can address these two states economic future.

Class Writing Assignment

Outline a strategic energy led economic development plan for the state of Michigan or Nevada that is one page in length, defining the economic challenge faced, goal, objectives, strategies, and tactics needed. This is just a starting point for a document that will grow.

Additional Readings

Bayerlin, Ulrich, *Sustainable Development* (Hamburg: Max Planck of Institute of Comparative and International Law, April 2009).

Brownfields and Land Revitalization, US Environmental Protection Agency, available at usepa.gov.

Campbell, Scott, "Green Cities, Just Cities? Urban Planning and the Contradictions of Sustainable Development," *Journal of the American Planning Association*, 62, 3 (Summer 2006): 296–313.

Carley, Sanya, Sara Lawrence, Adrienne Brown, Andrew Nourafshan, and Elinor Benami, "Energy-Based Economic Development," *Renewable and Sustainable Energy Reviews*, 15 (2011): 282–295.

Economic Impact Study of Ohio's Utica Shale, sponsored by the Ohio Chamber of Commerce, retrieved from http://urban.csuohio.edu/publications/center/center_for_economic_development/Ec_Impact_Ohio_Utica_Shale_2012.pdf.

Environmental, Socioeconomic and Governance Report, 2011, Goldman Sachs, retrieved from www.goldmansachs.com.

Hula, Richard and Rebecca Bromley-Trujillo, "Cleaning Up the Mess: The Redevelopment of Urban Brownfield's," *Economic Development Quarterly*, 23, 3 (August 2010): 276–287.

McMahon, Edward, "The Greening of the Real Estate Industry," *Urban Land Management*, The Urban Land Institute, January 20, 2012.

"North Dakota Sees Increase in Real GDP per Capita following Bakken Production," *Today in Energy*, US Energy Information Administration, July 12, 2013.

Rose, Adam and Dan Wei, "Macroeconomic Impacts of the Florida Energy and Climate Change Action Plan," *Climate Policy*, University of Southern California, 12 (2012): 50–69.

Soubbotuna, Tatyana, *Beyond Economic Growth, an Introduction to Sustainable Development*, 2nd ed. (Washington, DC: World Bank Group, 2004).

Urban Brownfields: Why Toxic Brownfields Are Tough to Redevelop, State Impact NPR, Texas, retrieved from www.npr.org/texas.

Vachon, Stephan and Fredric Menz, "The Role of Social, Political and Economic Interests in Promoting State Green Energy Policies," *Environmental Science & Policy*, 9 (2006): 652–662.

Yusuf, Juita-Elana (Wie) and Katharine A. Neill, "State Energy-Based Economic Development Policies and Examples," *Economic Development Quarterly*, 27, 3 (2013): 240–249.

Chapter 2

Economic Development and Energy Policy

Chapter Goals

1. Understand America's economic history.
2. Be prepared to discuss how America's economic history impact's today's economy.
3. Recognize the connection between energy policy and economic development.
4. Understand how state and federal government regulation of utilities influences economic development.
5. Be ready to discuss how America's transportation policy impact energy and economic development.

American economic development is tied to the energy policy. Economic transitions from an agrarian to an industrial, to today's technology and service-based economy are all tied to energy. Energy fueled all these transitions. The economic boom of the 1920s was not fueled by gin in a speakeasy. It was fueled by electricity and the new applications that improved productivity and innovation in America's growing manufacturing industry. The digital age and the explosion of computer technology were not only dependent upon the speed of a broadband network, but the availability of affordable and reliable power.

America's path toward a national economic development policy started with the founding of the colonies by Europeans searching for religious freedom and economic success. It continued on through the New Deal, to battles with inflation, into today's economic development era dominated by regional and state actors. America's energy policy exploded on to the scene with the Industrial Revolution that grew massive utilities to serve new factories. Government moved to regulate the energy industry to protect the public and provide an affordable and efficient power system for electricity and natural gas. The transportation industry was served by a growing network of competitors created by the breaking up of Standard Oil. The nuclear age entered and created new opportunities for "cheap" power.

All was well until the 1970s. The Middle East shocked the American economy with an oil embargo. A nuclear accident at Three Mile Island in Pennsylvania nearly killed the nuclear power industry. Environmental regulations began to assault the coal the nation depended upon for cheap electricity. Matched with the decline of the American manufacturing industry, America struggled to create an energy policy from the 1970s through the 1990s.

As the twenty-first century arrived, the connection between energy policy and economic development became clearer. Technological innovations began to make alternative energy more affordable. Technological innovations also enabled the shale gas explosion, which is accomplishing the dream of energy independence. The twenty-first century is proving to be the American Energy Age.

Economic Development Is about Wealth Creation for a Community's Residents

Economic development is the use of public subsidies to promote wealth creation for the larger community and not just the property owners involved. Economic development is the conduct of government and its private sector partners to spur wealth creation for its people. Economic development strategies are implemented at the state and regional levels. Regional and state economic development strategies are either basic, Building Block, or more advanced, proactive strategies known as the Five Drivers of Economic Development. All regions and states implement Building Block economic development strategies, preparing sites for development through land use planning, annexation, and eminent domain. Building Block strategies also address infrastructure development, workforce, tax policy, and quality of life issues. These fundamental economic development strategies are the floor and not the ceiling. They cannot be overlooked. Larger, pro-active efforts to retain and attract high-wage jobs center around the use of basic

Building Block economic development strategies but are targeted toward high-wage industry clusters.

The History of Economic Development Policy

It is challenging to trace the start of economic development policy. The creation of economic development incentives can be traced back to Colonial times in the United States. America was formed as a colonial empire for the unpopular religious sects of several countries in Europe, but was also created to take advantage of the new land's vast resources for the mother countries. The Massachusetts Bay Company was granted a monopoly on providing steel from the colonies for the site of North America's first steel factory.[1] Also, many of the legal requirements that impact economic development policy today were debated and included in the US Constitution during the Constitutional Convention in the 1780s. From eminent domain to the interstate commerce clause, the powers of the American government were created well over 200 years ago, and they ultimately still impact economic development in the twenty-first century.

In the 1890s and early years of the twentieth century, cities become the lifeblood of American society. From 1860 to 1910, the population of American cities went from 3.1 million to 9.1 million.[2] By 1910, 46 percent of Americans lived in cities with populations of over 2,500.[3] Growing cities struggled to address crime, poverty, and the host of urban issues still surrounding our cities. The population explosion of America's cities was tied to the growing dominance of American industrial might. This was the age of Rockefeller, Morgan, Frick, Carnegie, and others who built massive industrial empires.

Reform Movement Takes Hold

Urban reformers of the nineteenth century launched the City Beautiful Movement. Daniel Burnham, a leader in the City Beautiful Movement, linked the movement to the era's Progressive Movement, adopted by politicians such as President Theodore Roosevelt.[4] The City Beautiful Movement was in many ways a "reform" of a city's landscape. These reforms could take the form of improving sanitary conditions, the missions of Jane Addams' Hull House in Chicago, but more likely lead to the creation of monuments, public spaces, and private buildings of grand stature.[5] The best examples of large-scale City Beautiful Movement projects are the 1893 Columbia

Exposition and World's Fair in Chicago and the 1901 Washington DC Plan that resulted in much of the Mall in our nation's capital.[6]

Reform did not stop with architecture. As America entered the twentieth century, the challenges of industrialization of the quality of life of the nation took center stage. The power of the industrialist was challenged. The post-Civil War Era dominated by Laisse Fair Republicans ended with the assassination of William McKinley. Government policy began to address matters such as the age of workers, safety of workplaces, and taxes paid by corporations. The Progressive Era had begun. Other than a "Return to Normalcy" in the 1920s, American economic policy was centered on managing the impact of the industrial process.

New Deal Is Launched

However, the government did not stop with building grand buildings and public spaces to address our economic concerns. Direct government intervention greatly increased under President Franklin Delano Roosevelt's New Deal. In an effort to address the economic challenges of the Great Depression, the federal government launched a massive expansion of regulation of business as well as the creation of a wide range of programs. The TVA not only created a massive electricity generation initiative and brought power to millions, but also was the first example of regional planning.[7] In 1934, the Housing Act was brought into force, and the Federal Housing Administration was created to ensure loans for construction, renovation, and the repair of homes.[8] The New Deal launched a massive federal government program to address unprecedented economic challenges.

Post-World War II Launch of the Suburbs

The return of soldiers and the hopes of preventing a second Great Depression kept the federal government involved with economic issues in Post-World War II America. Not all of this involvement proved successful. The National Housing Act not only assisted the availability of low-cost housing, but it also enabled millions of Americans to leave the urban centers for the land of plenty—the suburbs.[9] The federal government provided assistance for millions to gain mortgages for new homes in the suburbs. In 1949, another Housing Act provided federal funds to cover the costs of purchasing blighted areas in the cities and permitted private developers to build houses there; but the result was disaster.[10]

Economic development became about demolishing America's urban core to build new structures. The enactment of the 1956 Federal Highway Act connected the nation, but resulted in the division of urban neighborhoods, displacement of urban residents, and the continued creation of massive incentives for the evacuation of the urban core to the suburbs.[11] Ironically, most urban mayors wanted the federal dollars for housing and expressways. Mayor Richard J. Daley in Chicago was seen as widely successful, in part because he was able to gain millions of dollars in funding for new housing projects and interstate highways for his city.[12]

America was booming in the 1950s. The United States was the only nation capable of rebuilding the world after World War II. Unfortunately, much of our well-intentioned federal laws and policies lead to the destruction of our urban cores in the process. The implementation of the federal housing and transportation laws were only the start of urban decline. The 1954 *Brown v. Board of Education* decision started the end of racial segregation in America's public schools. Over the next 20 years, as Brown was implemented and the busing of students across cities began, "white flight" hit America's urban areas. With cheap mortgages and highways built to reduce commute times, the loss of neighborhood schools and fears based upon racism put the evacuation of our urban cities on overdrive.

The 1950s also saw billions of dollars of funding pouring out of Washington to build a national highway system that President Eisenhower wanted for national defense purposes. Congress wanted them because it would be a major source of jobs for decades to come, building thousands upon thousands of miles of roads and bridges. Infrastructure development became a national economic development priority. The nation was at its peak of global power following World War II. The United States was building products for the world and, certainly, the government could come up with the funding to pay for it.

Transportation Revolution

Why did every Mayor want a freeway? First, the car had become the American dream. Americans owe the invention of the car to a German. Nicolaus Otto built the first practical combustion engine in 1876.[13] Henry Ford created his now famous assembly line process and brought the Model T to the masses.[14] His $850 Model T sold 10,607 units in 1908.[15] By 1916, with a fully operational assembly line, the price of the Model T dropped to $360 and Ford sold over 730,000 cars.[16]

America was not just enjoying the economic fruits of the Industrial Revolution, but was undergoing a complete "transportation revolution."

Mayors wanted freeways because the car had become "The Mode" of transportation. The era of trolley cars and rail had ended for cities because most residents could afford a car. In addition, mayors wanted the federal interstate highways to strengthen access to the Downtown portion of the city. The highways could serve as a link to growing suburban communities and bring workers Downtown.

However, the interstate highways in many cases killed neighborhoods by slicing them in half as well as disrupting previously operating commercial corridors. The traffic flow changed. The quality of the housing and neighborhoods dropped as a big, noisy freeway became a next door neighbor. The expressway didn't just serve to bring workers Downtown, it expedited the process of workers moving further and further out into the suburbs and contributed to the evacuation of many central cities.

The Collapse of America's Urban Centers

Again, the federal government was not standing idly by. Another activist President, Lyndon Johnson, launched his 1960s version of the New Deal known as the Great Society. Johnson's War on Poverty was closely affiliated with a racial justice program and centered on federal involvement to provide equal opportunity with an aggressive new initiation targeted at urban centers. The Office of Economic Opportunity was created in 1964 and coordinated overall efforts to implement a Job Corps in urban centers, Volunteers in Service to America (a domestic version of the Peace Corps), and allowed for and funded Community Redevelopment Corporations across the United States.[17] The Economic Development Act of 1965 was created to lead the nation's economic development efforts and the Department of Housing and Urban Development assisted in creating the Model Cities Program in 1966 to improve the coordination of existing urban programs and provide funding to local planning initiatives and local community organizations in the social services world.[18] Crime became a major concern in our urban centers. The 1965 Watts riot in Los Angeles was closely followed by urban riots in Detroit, Cleveland, Newark, Washington, and Cincinnati. Many of these urban cores have simply not recovered since.[19]

The Nixon Administration also launched an aggressive program for economic development. Besides taking an aggressive stance against forced busing of students to achieve racial desegregation, the Nixon Administration created the Revenue Sharing Program that attempted to get the federal government out of the business of many of these local development programs, but provided funding to the states for such efforts. The Nixon Administration launched wage and price control policies to attempt to address staggering inflation. The American steel and auto industry began to lose its post-World War II dominance in the 1970s.

The 1970s saw the birth of the Community Development Block Grants as a tool to provide funding, with few strings attached to help local and state governments address urban challenges.[20] In 1977, the Carter administration created the Urban Development Action Grant (UDAG) program to give distressed communities funds for residential or nonresidential use to create public–private partnerships in the urban core.[21] The first 11 years of the program saw over $5 billion in federal investment from this program, and successful projects such as Chicago's Navy Pier and New York's Battery Park have ties to UDAG funding.[22]

1970s and 1980s again saw larger events swallow efforts by the federal government to solve economic development challenges. These events involved the economic collapse of the manufacturing based economy. The 1970s also saw the birth of new environmental protection laws that create protections for our air, water, and land. The US Environmental Protection Agency and its state government counterparts across America would clearly protect our environment, but they created a whole new array of regulations and restrictions on our manufacturing based economy.

New York City and many others lost their manufacturing based jobs, and the economic impact on the cities and its people was substantial.[23] Even the federal government's efforts to provide $100,000,000 in loan guarantees to the once all-powerful American steel industry proved unsuccessful in protecting this industry and others against lower cost and high quality global competition.[24]

Welcome to the Global Economy

As the 1970s took hold, the impact of global economy was hitting the United States. America became a high-cost market with highly paid workers trying to keep up with inflation. These urban markets were often in the industrial heartland, in states such as Ohio—home to eight US Presidents, mostly during the zenith of the Industrial Revolution. These "rust belt states" were hit hard by economic decline. From 1979 to 1985, employment in manufacturing declined by 21 percent, with steel-related industries representing 40 percent of those loses.[25] From 1974 to 1999, the actual workforce in the steel industry decreased from 521,000 to 153,000.[26] Urban centers such as Cleveland, Youngstown, Pittsburgh, Wheeling, and others who were heavily concentrated in steel industry jobs were, and still are, devastated by these losses. The Midwest was not the only region to struggle. New York City declared bankruptcy and manufacturing began to leave high-cost markers in the Northeast and West Coast.

The government's reaction to these economic challenges was an effort to reduce overall taxes and regulation as well as to diversify regional economics by targeting investment in research and development and other high-tech initiatives. States such as Pennsylvania created the Ben Franklin Technology Partnership in 1982 to build a technology based economy by encouraging public–private partnerships with universities and the private sector.[27]

By 1994, states such as Ohio, Indiana, Connecticut, and others, as well as the federal government, invested $3,000,000,000 in such technology-based public–private partnerships.[28] However, developing these technology-based projects in the urban environments is not easy and does not involve the substantial number of jobs for workers without a four-year or advanced college degree that the manufacturing industry did.

Tax Incentives Everywhere

The 1980s also saw the birth of the Enterprise Zone to stimulate investments and employment.[29] British planner Peter Hall originated this idea as a means to create business growth in urban centers by providing tax incentives and reduced government regulation to businesses.[30] Enterprise Zone programs now exist in all 50 states, but are often available to suburban and rural counterparts, not just urban areas. In addition, state and local governments in the 1980s began the dramatic expansion of public finance tools to assist in the development of needed infrastructure targeted at specific development projects.[31] Matched by federal tax-exempt bonds, states also began offering a variety of industrial revenue bonds to finance buildings and equipment for the private sector.[32]

Information Age

The 1990s saw the explosion of the Information Age. The computer began not only to impact the operation of corporate America, but to impact the average consumer. Phones became mobile. The Internet connected the world and brought information once stored in libraries to everyone's desktops. Spurred by the innovation of technology-based companies in

Northern California, Silicon Valley became a region of economic importance, on par with Wall Street. Tech companies exploded. The government's economic policy was simply to stay out of the way and encourage the development of as much broadband as possible.

The twenty-first century is the Age of Energy. Technological innovations have reduced the cost of alternative energy production as well as spurred an explosion of domestic oil and natural gas production from once-dormant shale deposits in many regions across the United States. This growth of the energy sector is making the dream of American energy independence a reality and primarily providing a low-cost option for natural gas. The federal government, in response to an economic crisis, pumped billions of dollars in funding into the alternative energy sector, to promote its use and development. However, political divisions continued to deny the United States a larger role for the federal government in reaching consensus on a national economic development policy.

History of American Energy Policy

The history of the American energy policy is directly tied to the uses demanded by the market with a stated goal of establishing a diversified portfolio of energy sources. The government quickly moved to establish regulated monopolies for electric and natural gas utility services, while the courts broke up Standard Oil's monopoly over oil, which in turn brought about competitive market forces, primarily in the transportation energy market (figure 2.1).

America began using wood, still a major source of biomass energy, as its prime energy source. As the nation entered the Industrial Revolution, and exploration techniques became more advanced, coal, petroleum, and natural gas became the dominant sources of energy. The dominance of petroleum as an energy source is tied to the growth of the automobile as the prime means of transportation in the United States. Petroleum fuels nearly 100 percent of the American auto fleet. The nuclear age began in the late 1940s, but nuclear energy is still far behind the three dominant fossil fuel sources—in terms of its usage.

The history of the American energy policy is a story that began with the reign of the free market, moved to the regulation of the marketplace to promote efficiency and protect consumers and now is moving back toward deregulation. It is also a story of massive subsidies to promote specific sources of energy. The federal government is the prime driver of the national energy policy with the states playing the role of implementing federal law.

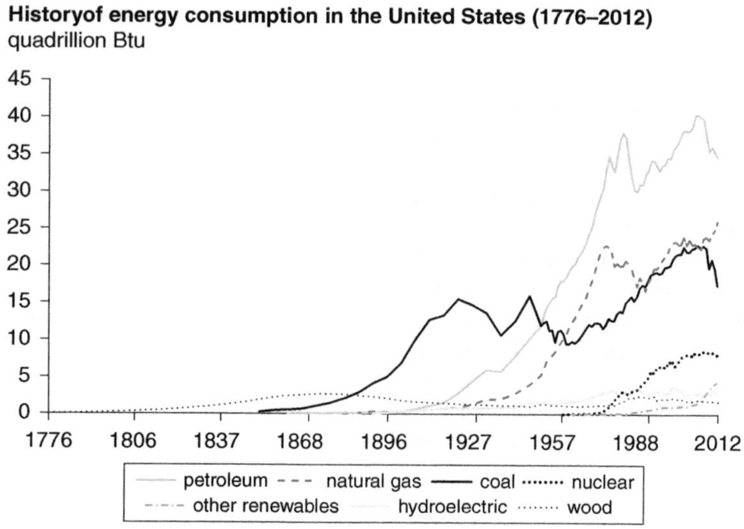

Figure 2.1 History of American energy consumption.

Source: US Energy Information Agency, retrieved from http://www.eia.gov/todayinenergy/detail.cfm?id=10, February 9, 2011.

Birth of the Natural Utility Monopoly

Thomas Edison's creation of the light bulb soon evolved into not just the development of electric lights replacing natural gas, but ignited the development of electricity operated manufacturing and processing, and consumer products powered by a newly developed electric power network. This late nineteenth century-early twentieth century technological innovation was impacted by its historical timing, from both a political and a business perspective. From a business perspective, the growth of electricity as a power source was perfectly time to expand the growth of the Industrial Revolution. Electricity's application to the nineteenth century factory enhanced productivity and profitability and created new industries along the way. Henry Ford's auto assembly plant ran much better with equipment power by electricity. From a policy standpoint, the electric power industry was born during the Progressive Era, when the public demanded greater regulation of business to soften the harsh edges of the nineteenth century factory.

Thus, early electric and natural gas services were quickly defined as "natural monopolies" and the public policy approach involved the government, not the market, regulating the operation of the electric and natural gas industry. A natural monopoly is a "firm or industry whose average cost per unit or production falls sharply over the entire range of its output."[33] Thus, a single firm, a monopoly, can supply the industry output more efficiently than can multiple firms.[34] The substantial capital and service cost of utility services, whether natural gas or electricity, made these industries natural candidates for monopoly regulations. To protect consumers and permit utilities to grow and prosper, state governments began to regulate electric and natural gas services as a natural monopoly service. Natural monopoly regulation permits government to award a monopolistic franchise to a specific company for a specific geographic region, but requires the company to provide a high-quality service at a specific price. The company in turn gains a guaranteed rate of return for its capital investment and operation of the business. The US Supreme Court in 1887 recognized the legality of regulating natural monopolies in this fashion and established the principal that state governments had the power to set rates for monopoly-style companies.[35]

Age of Utility Regulation

Following the market expansion of utilities and sanction from the US Supreme Court, the regulation of utilities began with municipalities and quickly spread to state governments. Municipal regulation of utilities began with the awarding of franchises to utilities to operate in the public right of way next to public streets.[36] Access to the right of way is essential for the delivery of electric and natural gas services to business and residential customers. It is the prime method of delivering their product. The late nineteenth century saw municipalities award multiple franchises to competing utilities.

In 1907, Wisconsin and New York took a broader approach to regulate the provision of utility services, through powerful state commissions that could permit municipalities to terminate existing utility franchises by buying the utilities assets, require a "certificate of public convenience and necessity" to control utility entry into the marketplace, establish utility rates, and regulate capacity additions and securities issuance of utilities.[37] The Wisconsin and New York laws led the way for 27 other states to follow suit between 1907 and 1914 by passing similar laws regulating the operation of utilities from a statewide perspective.[38] The Courts quickly agreed with municipal and state regulation of electric and natural gas services.

Market transformations also influenced the operation of the power industry during regulated times. Power sources started with wood and moved to coal, then expanded to oil and natural gas, eventually including nuclear—and now renewable—sources of energy again. Energy policy is driven by the technological innovations that discover new sources of power. The early 1900s saw coal become king due to its use in the industrial process. Coal production peaked in 1918 when it hit 678,000,000 tons of production.[39] The federal government, out of concern for this growing market control within the oil industry, adopted the Hepburn Act, which provided federal regulation to the oil pipeline industry.[40]

However, as America entered the 1920s, oil and natural gas began to play a more dominant role in the nation's energy production, as a "cleaner" fossil fuel. Oil constituted nearly one-fifth of the domestic energy market by 1925.[41] The increased reliance on oil in the 1920s is also related to the expansion of the automobile as the prime means of transportation. Oil also benefited from a lack of structured regulation in the same way the use of coal and natural gas was. In 1900, Standard Oil controlled 90 percent of the US market and, even after the breaking up of the company through the federal government's anti-trust lawsuit, Standard Oil still controlled 64 percent of the market in 1911.[42] The anti-trust litigation as a policy weapon failed to completely loosen thriving competition. Thirty-two firms in 1919 controlled 60 percent of oil production and in 1930, the thirty largest oil companies controlled 72 percent of the nation's oil refining capacity.[43]

Early twentieth century energy policy was driven by a massive new supply of diverse energy sources including oil, natural gas, and coal, with an industry moving from regional to national in focus serving a new and expanding industrial and transportation marketplace but resulting in a concentration of the energy industry around a handful of firms.[44] It is even more interesting to note that policymakers did not coordinate their treatment of different sources of energy or take a comprehensive regulatory perspective.

The federal government's first foray into energy policy was in the 1920s, with the passage of the Federal Water Power Act. This federal law created the Federal Power Commission, whose purpose was to license, permit, and encourage the development of hydroelectric power projects.[45] Around the same time, the federal government jumped into the oil industry by creating the Federal Oil Conservation Board to regulate the industry to address waste, declining reserves, and price instability.[46] The Federal Oil Conservation Board promoted the oil depletion allowance and started the process of opening up the substantial oil reserves located on federal land.[47] This permitted large firms to control production and reduce the amount of oil on the market, working to the benefit of the major oil companies.[48]

Thus, by 1929, 21 major oil companies controlled 60 percent of production, only 10 firms controlled the same percentage of refining, and fourteen companies controlled 70 percent of the pipelines.[49] As the nation entered the Great Depression, the other players in the energy industry were equally dominated by just a few companies. Eight holding companies controlled 85 percent of the natural gas industry and 22 electricity companies generated 61 percent of the electricity.[50] Seventeen bituminous coal companies controlled only 20 percent of the market, but eight anthracite coal companies controlled 70 percent of that market.[51]

Policymakers facing skyrocketing unemployment and a collapsed stock market were also faced with a power industry dominated by just a few powerful players. Again, policymakers reacted by empowering government to more aggressively regulate the utilities industry. In 1935, the Public Utility Holding Company Act of 1935 regulated the size and defined the geographic service area of electric and natural gas utilities and required all public utility companies to register with the newly created Securities and Exchange Commission to address questionable business practices.[52] The Federal Power Commission's jurisdiction was expanded to include all interstate electricity transmissions.[53] Next, the New Deal Congress expanded electric service to previously underserved rural markets with the passage of the Rural Electrification Act. This law provides federal loans to utilities to build electric distribution networks to rural communities through the creation of electric cooperatives who still operate in rural areas today.[54] The Natural Gas Act of 1938 gave federal government regulators the authority to set "just and reasonable" rates for interstate transmission and sales of natural gas, power to approve facility siting, and abandonment of facilities.[55]

The New Deal did not leave oil alone either. The Connolly Hot Oil Act and the Petroleum Allocation Board attempted to regulate oil production but ultimately failed, as additional oil discovery flooded the market with cheap oil.[56] Again, the result was continued market concentration, with 20 oil companies controlling 70 percent of the reserves and 76 percent of the refining capacity.[57] Even the New Deal could not control Big Oil. The federal government had equally big challenges regulating the coal industry during this same timeframe. However, coal production drove prices down, employment levels dropped due to technological innovations, and federal policies were more successful in creating safe working conditions than in regulating prices and production.[58]

Following the New Deal, World War II shaped federal energy policy for the next several decades. America entered the atomic age and federal policymakers followed suit. In 1946, the Atomic Energy Act was passed and the Congress determined that nuclear weapon development and nuclear power management would be under civilian—not military—control, through the creation of the Atomic Energy Commission.[59] This law also created a strict ban on the release of atomic technology to other world powers.[60] In 1954, Congress amended this act to give more flexibility for the exchange of information about atomic energy with friendly foreign powers and private companies in the growing nuclear power industry.[61] Congress recognized the atom could be more than a destructive force: that it could also be harnessed to create a new power source.

The Age of the Energy Crisis

All was fine in the energy sector as the United States dominated the global economy as the only world power left with manufacturing, distribution, and production capabilities following the devastation of World War II. Regulated power markets provided reliable power to the manufacturing industry. Detroit made increasingly bigger cars and more and more Americans could afford automobiles, turning America into a net oil importer, not exporter. Unfortunately for the United States, at about the same time, the global manufacturing industry caught up with the United States, global powers figured out that America was dependent upon them for energy. The Arab oil embargo of the early 1970s brought on by US foreign policy decisions to support Israel acted as an economic shockwave on the American economy. An oil industry still dominated by just a few players lacked supply, and prices skyrocketed. Of equally important concern was an unanticipated shortage of natural gas just a few years later. To make things even more interesting, the nuclear power industry barely survived a deathblow, with the near catastrophe at Three Mile Island nuclear plant in Pennsylvania.

These three solid punches to the energy industry, matched with a growth in industry and government spending in the 1960s, created widespread inflation and high-unemployment. This economic shockwave also ushered in a new era of pro-active federal energy policy. In 1974, the federal Energy Reorganization Act provided protection of nuclear workers, established the Nuclear Regulatory Commission, and split the responsibility for nuclear weapons and civilian nuclear power between the Department of Energy and the Nuclear Regulatory Commission respectively.[62] Responding to the skyrocketing oil prices, the Congress passed the Energy Policy and Conservation Act in 1975. This act established the Strategic Petroleum Reserve to ensure that the United States had a stockpile of oil in reserve in case of a national emergency, took the unusual step of establishing vehicle fuel economy standards, and extended oil price controls through 1979.[63]

1977 saw the passage of the Department of Energy Organization Act that created the Department of Energy as a Cabinet level organization, and the Federal Energy Regulatory Commission within this department to regulate energy issues.[64] The federal government also began the process of diversifying the nation's approach to energy used for transportation purposes. The National Energy Act of 1978 created tax incentives and disincentives, alternative fuel programs, energy efficiency initiatives, and other regulatory and market-based initiatives in response to the oil crisis. It developed a Gas Guzzler Tax for vehicles with mileage below a specified level, created income tax credit for the use of home solar, wind, or geothermal energy systems, set up wellhead pricing maximums and rules for high-cost gas to industrial consumers, and gave the Federal Energy Regulatory Commission jurisdiction over most natural gas production.[65] The National Energy Conservation Policy Act replaced Minimum Energy Performance Standards created just four year before and moved the United States from voluntary to mandatory energy standards.[66] Federal agencies were also required to do energy audits of their operations, the federal government became a source for families to gain loans for solar heating or cooling systems and schools, hospitals, local governments, and public housing authorities were provided grants for energy conservation.[67]

In response to the shortage of natural gas, the Congress enacted the Power Plant and Industrial Fuel Use Act in 1978 to restrict construction of power plants fueled primarily by oil or natural gas, and instead encouraged power plants fueled by coal, nuclear, and alternative fuels.[68] Coal again became popular. Again, in line with responding to the natural gas shortage, the Public Utility Regulatory Policies Act promoted not only the use of renewable energy, but encouraged cogeneration plants with the implementation left to state governments.[69]

1980 saw the passage of the substantial Energy Security Act comprised of the eight specific acts geared toward the continued diversification of the American energy marketplace.[70] The Synthetic Fuels Corporation was created to partner with industry to develop domestically-produced synthetic liquid fuels through a public-private-partnership built around research and development.[71] Loan guarantees were created for small-scale biomass energy projects, and other incentives were provided for the use of renewable energy sources such as solar energy and geothermal projects.[72] Energy targets to imports and consumption were set but only as goals and not mandates.[73]

Energy Deregulation

The policy response to the Great Depression and the Energy Crisis of the 1970s was simply more regulation and a new batch of incentives to promote additional energy diversification. In essence, the federal government promoted policies to move America away from a dependence on foreign oil. The federal government also attempted to address the consumption of energy by regulating the average miles per gallon a car could consume. Keeping the Progressive Era's monopoly regulation for utility operation was not questioned. Bringing in more competitive sources for power was not on the agenda. In an age of activist government, policymakers pushed the federal government to solve the energy challenge.

The election of Ronald Reagan as President brought a different perspective to the energy policy. Elected with a clear message that government was the problem not the solution, a conservative agenda dominated the Executive Branch. However, Democratic control of the Congress for the most part did not undo most of the 1970s era federal energy policy. However, the seeds of competition as an energy policy model were laid. In 1992, the National Energy Conservation Policy Act of 1978 was amended to create a framework for wholesale electricity generation, promote greater energy efficiency as a tool to address power costs and develop electric cars, and to continue down the energy diversification highway.[74] This act created a new category of electricity producer, the exempt wholesale generator, not subject to the constraints on nonutility electricity generation in the Public Utility Holding Company Act.[75] The federal government, for the first time in nearly 100 years, was searching for a way to make the market forces impact the utility industry.[76] These efforts were the beginning of the introduction to competitive electric and natural gas services. Nearly half the states now offer competitive retail electric and natural gas services while the federal government maintains regulatory control over its production and distribution.

As America entered the twenty-first century, federal policymakers continued with efforts to diversify sources of energy by subsidizing renewable sources of energy, in the hopes of developing a market share for the companies providing such services. The federal stimulus package created billions in funding for the renewable energy industry, but many of these industries, such as solar, struggle to compete with lower cost global competitors. America's energy policy is focused on assuring abundant supplies at reasonable prices while limiting the market power of major companies and developing inter-fuel competition through a federal-state regulatory framework.[77]

Nevada and Michigan's Energy Led Economic Development Goals and Measures of Success

Nevada and Michigan have both set goals of diversifying their economy, and gaining jobs in the energy sector is a part of that strategy. Nevada is focused aggressively on renewables, and Michigan is strong in that area as well. Detroit needs to find goals that benefit from its auto manufacturing industry but do not compound the problem by recruiting more auto manufacturers. Detroit needs to embrace globalism, technology, and advanced services to complement their manufacturing strength, which they cannot give up. Nevada lacks a manufacturing past, so it need to not set goals that include manufacturing, but instead look for technology, advanced services, and global companies.

Some early victories for Nevada and Michigan's energy led economic development strategies provide insight into what the overall objectives should be for these states' strategic energy led economic development plan. Nevada retains and attracts energy companies. 1,300 jobs were added to Nevada's clean energy sector in the third quarter of 2014.[78] Jobs in this sector are at 24,600, and this reflects a growth of 2,400 jobs in one year paying over $64,000 annually in wages.[79] Even Las Vegas has higher demand for urban housing, as urban sprawl creates traffic congestion too big of an issue for many residents to face.[80] Nevada announced the decision by electric carmaker Tesla to open a $5,000,000,000 electric battery plant in Reno.[81] Nevada provided a $1,300,000,000 economic development incentive package that was twice what Tesla asked for, and is controversial, as many question the economic return on investment even if all these 6,500 jobs appear.[82] However, Tesla's Nevada investment still marks a major energy led economic development milestone for a state desperately looking to diversify its economy beyond gaming and tourism.

The Nevada solar industry is a bright spot for the state's energy led economic development policy. 161 solar companies employing 3,100 people call Nevada home.[83] In 2013, there was 47 MW, constituting a $108,000,000 investment in solar electric capacity installed, meaning the state is ranked twelfth in the nation, and Nevada overall ranks 6th in the nation, for solar capacity.[84] Prices for solar installation have fallen 24 percent in the last year—far above the national average of an 8 percent decrease.[85] Nevada is the site of several recent utility-scale solar array developments including at Copper Mountain 2 in Boulder City, Moapa Solar Project in Clark County, and the Crescent Dunes Solar Energy Project in Nye County.[86] These projects alone can power over 100,000 homes.[87] More solar investments are clearly possible. The Department of the Interior recently determined that Las Vegas could create 6,700 MW more solar energy, which is enough to power over 6,700,000 homes.[88] Not all of Nevada's solar industry news is good. Solar manufacturing continues to struggle in the state, as it does across the United States, as China's massive subsidies to this industry harm American solar manufacturing efforts. In fact, a Californian solar manufacturer who gained millions of dollars from the federal stimulus package closed its Las Vegas solar facility after just one year.[89]

Michigan is scoring some early victories with their energy led economic development strategy. Michigan is home to about 120 companies that supply wind components and employ 4,000 in the state.[90] Michigan's renewable energy capital investment from 2009 to 2013 was over $2,000,000,000, with more than 1 GW of new wind capacity installed.[91] Renewable energy manufacturing generated $4,900,000,000 in economic output annually.[92] Nearly 70,000 Michigan workers are employed in the energy efficiency and renewable energy industries.[93] Fuel cells, energy-efficient lighting, solar photovoltaic, electric car battery technologies, and wind turbine manufacturers have seen the most growth.[94] No doubt, Michigan has more work to do on its economy, but the state's energy led economic development strategy is beginning to provide some economic traction for a once great industrial state.

Class Debate

The class will continue their exercise to support the creation of a strategic energy led economic development plan for Michigan or Nevada. Class discussion will focus on how energy led economic development can address these two states' economic futures.

Class Writing Assignment

Draft the goal and three measurable objectives for the Michigan or Nevada strategic energy led economic development plan.

Big Questions

1. Does the "carrot" or the "stick" work better as the government's energy policy?
2. How does the government regulate the power industry?
3. Is the market a better answer for the energy industry?
4. Can the federal government truly change the behavior of the energy market?
5. How have the interstate highway systems run counter to energy conversation measures?

Additional Readings

Clark, John G., *Energy and the Federal Government: Fossil Fuel Policies, 1900–1946* (Urbana and Chicago: University of Illinois Press, 1987).
Hirsh, Richard F., *Power Loss: The Origins of Deregulation and Restructuring in the American Electric Power Industry* (Cambridge, MA: MIT Press, 2000).
Lifset, Robert (ed.), *Energy in Historical Perspective: American Energy Policy in the 1970s* (Norman: University of Oklahoma Press, (2014).
MacAvoy, Paul, *The Natural Gas Market: Sixty Years of Regulation and Deregulation* (New Haven, CT: Yale University Press, 2000).
Melosi, Martin, *Coping with Abundance: Energy and Environment in Industrial America* (Philadelphia: Temple University Press, 1985).
Pope, Daniel, *Nuclear Implosions: The Rise and Fall of the Washington Public Power Supply System* (New York: Cambridge University Press, 2008).
Vietor, Richard H. K., *Energy Policy in America Since 1945: A Study of Business-Government Relations* (Cambridge: Cambridge University Press, 1984).
The White House, *Energy and Environment,* available at http://www.whitehouse .gov/issues/energy-and-environment.
Yergin, Daniel, *The Prize: The Epic Quest for Oil, Money and Power* (New York: Simon and Schuster, 1991).
Yergin, Daniel, *The Quest: Energy, Security, and the Remaking of the Modern World* (New York: Penguin, September 2011).

Chapter 3

Energy Supply and Demand

Chapter Goals

1. Understand what energy is.
2. Recognize what types of energy Americans consume.
3. Understand the connection between the supply and demand for different energy sources.
4. Be able to define the different sources of energy that supply the American economy.
5. Be aware of positives and negatives of fossil fuel and renewable energy sources.

The sun is of the most valuable of Phoenix's assets. The sun is a major source of energy for the planet and solar energy. First Solar found the allure of the Phoenix sun too much to ignore. First Solar started in Toledo, Ohio—the "Glass City"—working on transforming itself into the "Solar City." First Solar is one of America's leading solar equipment manufacturers. A strong solar research and development initiative led to the creation of First Solar in Toledo. It was not enough to keep the First Solar headquarters away from the Arizona sun. The company followed its product's energy supply for its corporate headquarters.

Regions with a supply of energy can utilize that asset to launch an industry initiative. Those regions with an opportunity for large-scale solar facilities can turn those projects into a campaign for the retention and attraction of corporate headquarters around this industry. The supply of energy a region possesses and the demand for that energy is an opportunity to develop an energy led economic development initiative. All forms of energy can be a regional economic stimulant. America's economic success depends upon the development of domestic sources of energy from renewable and nonrenewable sources. The plentiful supply of energy provides a region with a low-cost source of power attractive to companies and an opportunity to retain and attract high-wage manufacturing, distribution, research and development, and service jobs connected to that energy supply.

What Is Energy?

Everything runs on energy. From people to machines, all devices on earth operate through the use of energy. Energy is either stored (potential) or working (kinetic). Energy supply comes in a variety of forms such as heat, light, motion, electrical, chemical, nuclear, and gravitational.[1] Nonrenewable fossil fuels of coal, natural gas, nuclear, and oil will remain the core of America's energy supply; however, renewable sources such as wind, solar, hydropower, geothermal, and biomass are gaining a larger share as the nation's energy source. The power sector will continue to rely on fossil fuel in the future, but its percentage of use will decline from 68 percent to 57 percent by 2035.

Overall demand for energy in the United States is expected to slow down, based upon the population and demographic changes coming; however, the industrial and commercial use of energy will remain strong. The global demand for energy will surge, particularly in emerging economies. Renewable and natural gas demand are expected to grow while demand for coal will drop in the United States and Europe but surge substantially in China, India, and other emerging markets.

Supply of Energy

Americans get their energy from a range of sources. However, the vast majority of the American energy supply comes from nonrenewable sources—primarily fossil fuels of oil, natural gas and coal. The domination of fossil fuels in the American energy supply has existed for over 100 years, with the advent of the automobile operating on oil based gasoline and the access to low-cost coal as the prime method to produce the electricity that powers factories and commercial centers—as part of the Industrial Revolution. Prior to the domination of fossil fuels, biomass, in the form of wood, was the dominant source of energy supply (figure 3.1).

The earth produces these sources of energy, which are all extracted, distributed, and processed into energy used by a majority of Americans. Nonrenewable energy sources constitute over 80 percent of all energy used in the United States.[2] Including nuclear energy in the fossil fuel mix, fossil fuels are the source for over 90 percent of all American energy output.[3]

Within the fossil fuel category, natural gas consumption is forecast to grow over the next thirty years, driven by the growth of domestic sources of natural gas from shale-based developments.[4] Energy efficiency mandates will drive down energy consumption for oil as the overwhelming provider of energy for America's auto driven society.[5] Coal consumption is scheduled to remain even at current levels, heading into 2040.[6]

Renewable energy sources constitute solar, wind, biomass, biofuel, hydropower, and geothermal. America gains just under 10 percent of their power from renewable sources.[7] Growth in renewable energy sources is expected heading into the future. The share of renewable energy sources, not including biofuels, is forecast to grow from 9 percent in 2011 to 13 percent of all American sources for energy.[8]

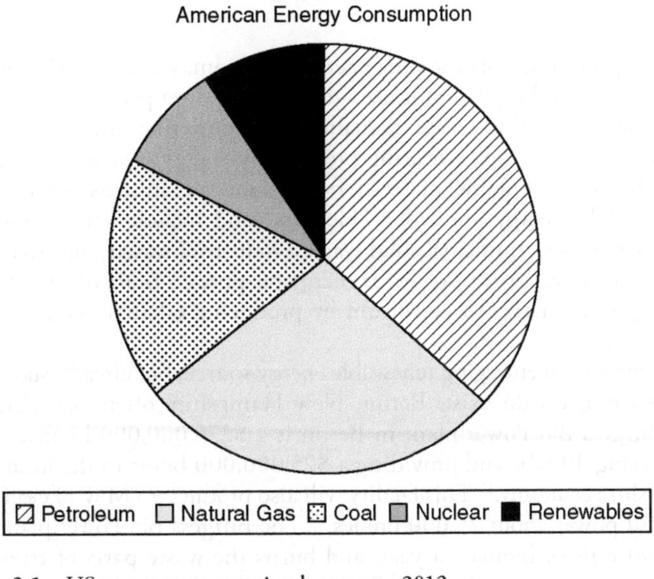

Figure 3.1 US energy consumption by source, 2012.

Source: US Energy Information Administration, Annual Energy Outlook 2010 with Projections to 2035.

Growth of the use of renewable energy sources is, in many cases, driven by federal and state regulations that create incentives and mandate the use of renewable sources for energy. The use of biofuels is expected to triple by 2035, based upon substantial domestic and global subsidies to move more automobiles off traditional oil and onto other sources of energy.[9]

The debate between renewable energy and nonrenewable energy advocates, regarding which source of power is more affordable, is fierce. Many elements impact costs, ranging from the cost of the energy source commodity to the costs of transporting the power to a plant for distribution. Wind and hydropower are generally the least expensive renewable energy source, and off-shore wind, solar, and biomass are generally more expensive than nonrenewable energy sources.[10] Nuclear power leads the list of high-cost nonrenewable sources of energy, with the shale expansion moving natural gas to the lowest end of the cost scale for all sources of power.[11] Geothermal is a lower cost source and coal is at present more expensive than natural gas and geothermal but still cheaper than off-shore wind, solar, biomass, and nuclear.[12]

Biomass

Plants capture and store the sun's energy, and biomass energy is the method used to harness this power for use—burned in power plants, fermented to make fuel, digested to create methane gas for powering turbines, and gasified through a complex heating process. Bioenergy is renewable energy made from any organic material from plants or animals sources, and sources of bioenergy are called "biomass" and include agricultural and forestry residues, municipal solid wastes, industrial wastes, and terrestrial and aquatic crops grown solely for energy purposes.[13] Biomass can play the role as an alternative to petroleum by producing ethanol, biodiesel, and electricity.[14]

Biomass is an emerging renewable energy source, but already-successful biomass projects do exist. Berlin, New Hampshire offers one example. The Burgess Bio Power plant in Berlin is a $275,000,000 biomass facility creating 40 jobs and providing a $25,000,000 boost to the local New Hampshire economy.[15] This facility will also produce 75 MW of electricity that will power about 75,000 homes.[16] The Burgess Bio Power plant uses 750,000 tons of biomass a year, and burns the waste parts of trees that cannot be turned into board.[17] Power from the plant is sold to the state's largest utility, Public Service of New Hampshire.[18] The facility is located on a former site of a paper factory site that closed in 2006 after a 100 years of operation.[19]

Sources of biomass are wide and located across the United States. Crop residues from wheat, corn, soybeans, cotton, barley, sorghum, barley, oats, rice, rye, canola, dry edible beans and peas, peanuts, potatoes, safflower, sunflowers, sugar cane, and flaxseed are all sources of biomass power.[20] Forest residues from logging trees and primary and secondary mill residues from manufacturing wood are also sources of biomass material.[21] Landfills are substantial sources of biomass energy, through the methane gas emissions produced by aging municipal solid waste.[22] Wastewater treatment plants also offer a source of methane gas that can be used as an energy source.

Developing crops specifically for energy are regarding as the largest opportunity for expanding the use of biomass. "Energy" crops are tall grasses like switch grass and miscanthus, and fast-growing trees like hybrid poplar and willow.[23] It is estimated that 400,000,000 dry tons of energy crops could be used by 2030.[24] Agriculture residues left as waste after a crop harvest could produce 155,000,000 dry tons of biomass for energy without impacting the current food crop market.[25] Household and yard waste and construction and demolition debris currently thrown away can provide 35,000,000 dry tons of biomass.[26] 60,000,000 tons of manure is produced in the United States and anaerobic digesters can be used to gain energy.[27] Logging waste could provide 20,000,000 of dry tons of biomass.[28] Biomass sources exist, and these sources can be harvested without impacting other markets.

Sources of biomass are available across the United States, but they do tend to be dominated by regions with substantial farm and agricultural sites. Major urban centers, through landfill and wastewater treatment plants, also have opportunities to use biomass as an energy source. Mountainous and desert regions are not the best sources of biomass (figure 3.2).

Biomass sources constitute nearly 700,000,000 dry tons by 2030, and there is enough biomass to produce 54,000,000,000 gallons of ethanol. This would constitute four times as much corn ethanol as the United States produced in 2010, or 732,000,000,000 kWh of electricity—that is 19 percent of total US power consumption.[29]

Advocates of biomass point to the substantial economic benefits, but the widespread adoption of biomass as a major power source struggles, based upon the cost of production. Numerous projects in California, Nevada, and other states have been closed due to the cost of energy production compared to other lower-cost methods.[30]

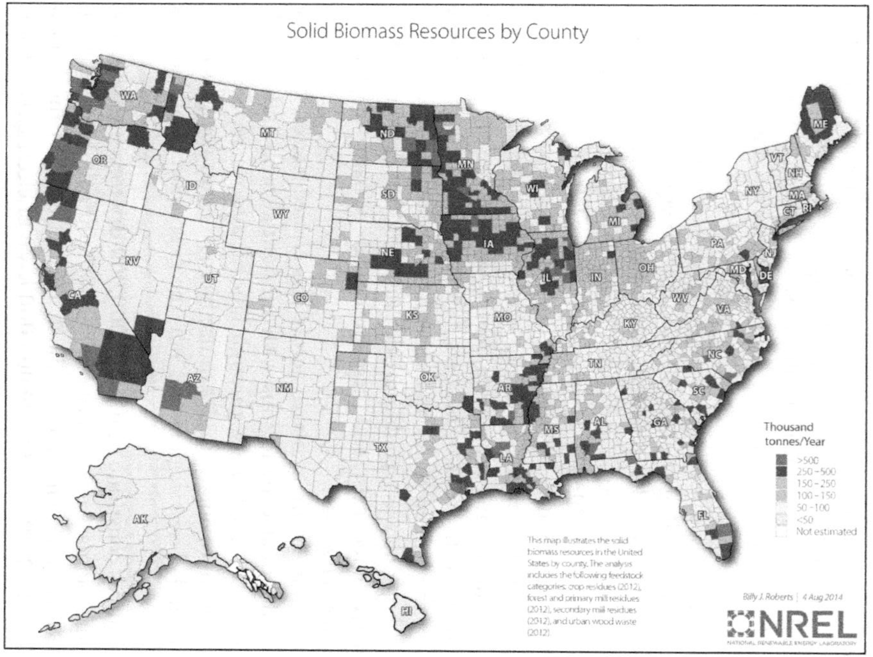

Figure 3.2 Biomass opportunities in the United States.

Source: This image has been reprinted with permission from the National Renewable Energy Laboratory (http://www.nrel.gov/gis/biomass.html). Accessed December 18, 2014.

Coal

Coal is a burnable carbonaceous rock formed from the remains of trees, ferns, and other plants that existed and died some 280–345 years ago.[31] Not all coal is alike. Coal may differ in sulfur, mercury, and heat energy content.[32] Coal is grouped into four categories based upon heat content: anthracite, bituminous, subbituminous, and lignite.[33] Anthracite has the best heat content and lignite the worst.[34] 90 percent of the US coalmines are bituminous.[35]

The United States is reliant on coal for much of our energy needs. Coal accounts for 95 percent of the nation's fossil energy reserves, nine out of every ten tons of coal used in the United States are used for electricity generation, and every American uses an average of 3.8 tons of coal each year.[36] While no one will claim that it is the cleanest version of energy around, coal is the United States'—and the world's—largest source of nonrenewable energy.

Many of the poorest, rural regions in the United States are the best sources of coal and are reliant on coalmining, distribution, and energy production for their economic livelihood. In fact, a 2001 study found that coal alone had a $10,000,000,000 economic impact on the states of Kentucky, West Virginia, and Pennsylvania, and employs thousands of workers directly and indirectly through a substantial supply chain.[37] Coal is also America and the world's largest and cheapest source of energy.

The United States holds the world's largest coal reserve.[38] America produces a billion short tons of coal, and 92 percent of this coal generates 37 percent of the nation's electricity in domestic power plants.[39] 1,400 coal-fired electricity generating units operate in 600 power plants.[40] Coal is mined across the United States and comes from 25 states.[41] Wyoming leads the nation in coalmining, followed by West Virginia, Kentucky, Pennsylvania, and Texas.[42] In fact, the Western states produce nearly half of the coal produced in the United States.[43]

The demand for coal is in decline due to federal clean air regulations and the explosion of new domestic natural gas and oil wells from previously dormant shale deposits. However, America is a net exporter of coal, and exports doubled from 2000 to 2012.[44] Coal exports include metallurgical coal for steel production and steam coal for electricity generation and industrial production.[45] Europe followed by Asia is the largest recipient of American coal.[46]

Coal has its opponents. The major argument against the use of coal is its environmental impact on the regions that mine it and the air people breathe near the coal burning power plants. Coal emits toxins such as sulfur dioxide, nitrogen oxide, mercury, arsenic, hydrogen chloride, and carbon dioxide.[47] Coal is seen as one of the major causes of global warming, as it produces 31 percent of the greenhouse gas carbon dioxide.[48]

Geothermal

Heat from the Earth can be and already is accessed by drilling for water or steam wells in a process similar to drilling for oil, and geothermal advocates see this energy source as an enormous, underused heat and power resource that is clean (emits little or no greenhouse gases), reliable (average system availability of 95 percent), and homegrown (making us less dependent on foreign oil).[49] Geothermal resources range from shallow ground to hot water and rock several miles below the Earth's surface, and even farther down to the extremely hot molten rock called magma.[50] Mile-or-more-deep wells can be drilled into underground reservoirs to tap steam and very hot water that can be brought to the surface for use in a variety of applications.

Case studies are everywhere about the efficiency and lower cost of geothermal heat pumps. Schools in Lincoln, Nebraska, used 26 percent less energy with a geothermal heat pump.[51] Fort Polk family housing power costs were reduced by 33 percent—with a 43 percent reduction in the peak summer demand—by geothermal heat pumps.[52]

Geothermal wells are drilled into an underground geothermal reservoir of water as hot as 700°F (371°C).[53] The trapped steam turns a turbine that produces electricity.[54] Ground source heat pumps move fluids through continuous pipeline loops connected to residential or commercial properties, and the pipeline loop pulls heat out of a building during the summer and returns cooler fluid to cool the building.[55]

In the United States, most geothermal reservoirs are located in the western states, Alaska, and Hawaii (figure 3.3).[56]

Part of the challenge for the widespread growth of geothermal energy as a major source of power is its location in less populated regions of the Western United States, where the demand for power is much lower than in the more densely populated regions of the country.

Hydro Energy

Hydropower, energy moving from water, is one of the oldest forms of energy production on the planet. It is a renewable source of energy created

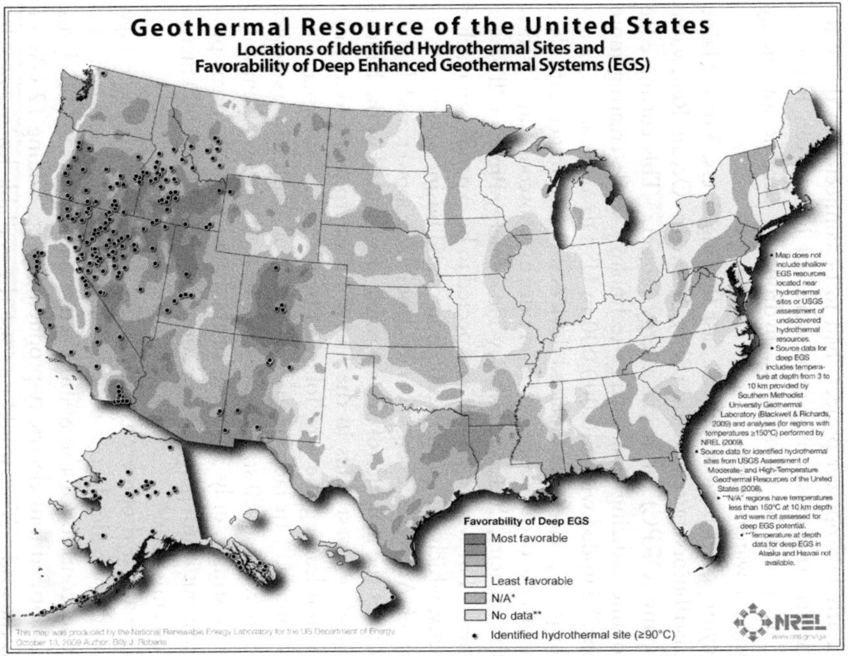

Figure 3.3 Geothermal resources of the United States.

Source: This image has been reprinted with permission from the National Renewable Energy Laboratory (http://www.nrel.gov/gis/images/geothermal_resource2009-final.jpg). Accessed December 18, 2014.

by rain and then by the harnessing of the force of the water in a pipe to turn turbines in dams, which of course generate electricity. The faster the water flow—think Niagara Falls—the more energy that can be harnessed. Waterpower was first used to produce energy in an industrial setting in 1880, when 16 brush-arc lamps were powered using a water turbine in a chair factory in Grand Rapids, Michigan.[57] The first hydroelectric plant in America was built in 1882 to power a factory and homes in Appleton, Wisconsin. Early hydropower projects had to be connected to the water source, as electric transmission systems were not invited yet. However, once electricity could be transmitted, hydropower became an energy source beyond the water's edge.

Not all hydropower projects involve fast flowing rivers. Three Maine electricity distributors purchase electricity from the Ocean Renewable Power Company (ORPC) for a tidal energy project.[58] This cutting-edge project is commercializing tidal energy technology while creating more than 100 jobs.[59] Funded with $10,000,000 in US Department of Energy resources, the pilot project phase placed devices designed to generate electricity from both the ocean's ebb and flood tides in Cobscook Bay.[60] When running at full capacity, the Cobscook Bay Tidal Energy Project will power 75–100 homes.[61] The Maine project is expected to be expanded to reach a 4 MW capacity, which will power 1,000 homes and businesses.[62]

7 percent of America's energy comes from hydropower and hydropower constitutes the majority of all renewable energy used in the United States.[63] The West Coast leads in the nation in waterpower production, with Washington leading the way, followed by Oregon, California, New York, and Montana.[64] In fact, 66 percent of all of Washington's power comes from water, but all 50 states generate some level of power from water sources.[65] Interestingly, only 3 percent of the nation's 80,000 dams produce electricity, but it still permits the nation to avoid 200,000,000 metric tons of carbon emissions.[66]

The market for hydropower is substantial. Only a small percentage of dams actually provide electric power. The vast majority are used as a means of flood control. Of the 80,000 active dams, over 50,000 dams could be transformed into electricity-producing dams, adding 12 GW of electricity to the grid.[67] This would increase the available hydropower by 15 percent.[68]

Much like other energy sources, not everyone is happy with hydropower. While not as bad as coal, the reservoir of water for hydroelectric power releases both carbon dioxide and methane from rotting plants and trees. Dams are also a problem due to the substantial disruption to the river habitat that they cause. Damns take nature's path for a river and redirect it to harness its energy. Damns are also expensive. While damns do pay for themselves over time, the upfront capital costs are substantial. American Municipal Power announced plans to retrofit five dams for electric production and the cost was $1,900,000,000.[69]

Nuclear Energy

Nuclear energy is another source of power for the United States. Atoms consist of a nucleus at its core surrounded by electrons. Protons and neutrons constitute the nucleus. Hydrogen, the lightest element, has one proton, uranium, the heaviest natural element. The nucleus of an atom is held together with tremendous force. When a nucleus splits, it can create great force. A nuclear power plant collides neutrons with uranium atoms and the split creates a chain reaction. Control rods contain the chain reaction but produce substantial amount of heat. This 520 degrees of heat spins turbines connected to generators and, thus, produce electricity.[70]

Nuclear energy is experiencing a mini-renaissance thanks to improving prospects for small reactors. Safety concerns following the Three Miles Island disaster in the 1970s stopped the growth of the American nuclear power plant. Cost is a concern for nuclear power. A traditional nuclear plant costs more than $10,000,000 to generate over 1,000 MW.[71] That sum could be half the revenues of a middle-sized electric utility company, and the risk is far greater than the reward at those numbers. However, based upon nuclear power plant innovations from military contracts, many companies are exploring smaller 200–300 MW nuclear facilities that only cost $2,000,000.[72] More technology advances are needed and the federal government provided a substantial research grant to find those answers; however, the small nuclear reaction may make nuclear power a bigger player in future American energy needs.

Nuclear energy is a highly efficient form of energy production. One pound of uranium has as much energy as 3,000,000 pounds of coal and is readily found in the Western United States, with Wyoming producing the most uranium.[73] The United States possesses 300,000,000 pounds of uranium reserves and uses about 40,000,000 each year.[74] Most uranium used in the United States is imported.[75] Mined uranium ore is processed in one of eight active processing plants to be turned into a useful fuel. There are eight active processing plants in the United States, although eight are inactive.[76]

Two-thirds of American nuclear reactors are pressurized water reactors and one-third are boiling water reactors.[77] Water boils into steam and is then sent through a turbine to produce electricity in a boiling water reactor.[78] Pressurized water reactors hold the core water under pressure and create the heat for energy production through a heat exchanger.[79] American nuclear power plants range in size from about 60 MW built in the 1960s to over 1,000 MW in more modern units.[80]

1951 was the first year an experimental reactor used uranium to generate electricity, but the first large scale commercial nuclear plant opened in Pennsylvania in 1957. The last new reactor began operation in Tennessee in 1996. American nuclear power generates 790,000,000,000 kWh from 104 commercial nuclear plants and provides more than 19 percent of the nation's electricity and 8 percent of all the energy consumed. 2012 saw the first new nuclear plant approved for construction in 30 years. The list of top states for nuclear energy starts with Illinois and is followed by Pennsylvania, South Carolina, New York, and North Carolina.

Nuclear energy probably wins the award for being the most controversial source of energy. Substantial environmental concerns exist about the danger of the energy process, but time proves more serious concerns exist related to how to handle the toxic waste by-products of the nuclear process. 70,000 metric tons of highly radioactive waste had been produced by American nuclear reactors.[81] In the mid-1970s, spent uranium was no longer permitted to be reprocessed into new fuel, as a by-product of this process is plutonium, which is a prime element of nuclear weapons. Since that time, nuclear waste is stored at existing nuclear plants, and consensus on the location for a single nuclear waste site have proved difficult.

Oil

Oil is a fossil fuel extracted from the earth and is the result of prehistoric organisms transforming into the black liquid that powers nearly all of America's cars. Oil is a nonrenewable fossil fuel with a high carbon content. A number of terms are used interchangeably in the oil footprint. Crude oil is a liquid hydrocarbon mixture found in underground reservoirs, and remains liquid on the surface. Petroleum products are produced from the processing of crude oil and the extraction of liquid hydrocarbons. Petroleum includes both crude oil and petroleum products.

America produces 5,400,000 million barrels of crude oil per day, 164,000,000 barrels per month, and almost 2,000,000 barrels per year.[82] The top five oil producing countries in the world are Saudi Arabia, Russia, the United States, Iran, and China, and the top five natural gas producing countries are United States, Russia, Canada, Algeria, and Iran.[83]

Once crude oil is found and removed from the ground, it is transported to an oil refinery. Crude oil is transformed into multiple products by a refining process that separates the different hydrocarbons to make petroleum gas for heating, cooking, and making plastics; gasoline; kerosene for jet and tractor fuels; diesel distillate for diesel fuel and heating oil; lubricating oil for motor oil, grease, and other lubricants; heavy gas for industrial fuel; and residuals, including coke, asphalt, tar, and waxes.[84]

The United States consumes 186,000,000 barrels per day (MMbd) of petroleum products and, for now, is the world's largest petroleum consumer. The United States imported 11.0 MMbd of crude oil and refined petroleum products in 2012, but due to the expansion of domestic oil sources in once-dormant shale deposits, oil imports are on the decline. World petroleum consumption is almost 84,000,000 barrels/day.[85] The United States imports oil from all over the world, not just one region, with the top five imports coming from Canada (2,100,000 barrels per day), Mexico (12,000,000 barrels per day), Nigeria (1,100,000 barrels per day), Saudi Arabia (1,000,000 barrels per day), and Venezuela (1,000,000 barrels per day).[86] In fact, the majority of America's oil imports are from the Western Hemisphere, and not the Middle East.

The decline in oil imports is a direct result of the discovery of a new supply of domestic oil from once-dormant shale deposits. Shale gas is trapped natural gas within sedimentary rocks that is extracted through high-technology horizontal drilling and hydraulic fracturing. Oil has detractors. As an energy source it is nonrenewable, the processing is fraught with environmental concerns, and its production and use is a major contribution to greenhouse gas emissions. However, oil remains the substance with which the vast majority of Americans drive to work every day.

Natural Gas

Natural gas is another nonrenewable fossil fuel, formed by buried plants and animals and exposed to intense heat and pressure. Carbon in natural gas preserves the energy these plants and animals had. Like oil, natural gas production into energy begins with the extraction, continues with treatment to remove impurities and transportation via pipeline to power plants, and ends with its combustion to generate electricity.

Natural gas is driving an American energy revolution in regions of the United States that, in the past, have not gained from the energy economy. North Dakota is one such example. North Dakota is "shale gas central," and its economy is the fastest growing in the United States. It is growing twice as fast as Texas, who is the second fastest growing state, and faster than even China. Natural gas' growth is also driving down the cost of energy and creating an opportunity to reshore American manufacturing jobs that are energy intensive.

Power plants burn the gas in a boiler to produce steam that a steam turbine uses to generate electricity, to burn the gas in a combustion turbine to generate electricity, or to burn the natural gas in a combustion turbine and use the exhaust to make steam to drive a steam turbine.[87]

Natural gas provides a substantial source of energy for heating America's homes, offices, and other business facilities. The United States is estimated to have enough natural gas to meet 100 percent of the current domestic demand for at least 90 years.[88] In fact, 95 percent of the natural gas consumed was domestic. The future looks even brighter for natural gas. American natural gas production is expected to increase from 23,000,000,000 trillion cubic feet in 2011 to 33,100,000,000 cubic feet in 2040, which constitutes a 44 percent increase.[89] Shale gas production is the cause of this planned increase, as it moves from 7,800,000,000 cubic feet in 2011 to 16,700,000,000 cubic feet in 2040.[90]

Hydraulic fracturing is a drilling technique in which water, chemicals, and sand are pumped into a well to unlock the natural gas long trapped in shale formations by opening cracks in the rock and allowing natural gas to flow out.[91] Horizontal drilling techniques enable shale gas to be extracted more efficiently.[92]

Many regions across the United States are benefiting from the shale gas expansion. Regions benefiting from the shale expansion not only include the traditional energy leader Texas, but also include North Dakota, Pennsylvania, Ohio, West Virginia, and other states in search of economic transformation (figure 3.4).

Natural gas, and shale gas in particular, have strong opponents. First, shale gas may be plentiful, but it will not last forever and is a nonrenewable source of energy. Also, concerns about shale gas wells impacting water quality exist, as most shale gas wells are deeper than traditional wells. Finally, there is much debate about what chemicals are used for the fracking process.

Solar

Solar energy uses the sun as fuel to create heat or electricity. Solar technologies are passive or active. Passive solar powers structures and active solar power uses heat, hot water, or electricity through Solar Thermal panels.[93] Solar energy is a renewable source of energy, but is an intermittent source because it is not always sunny. Clouds and rain impact the availability

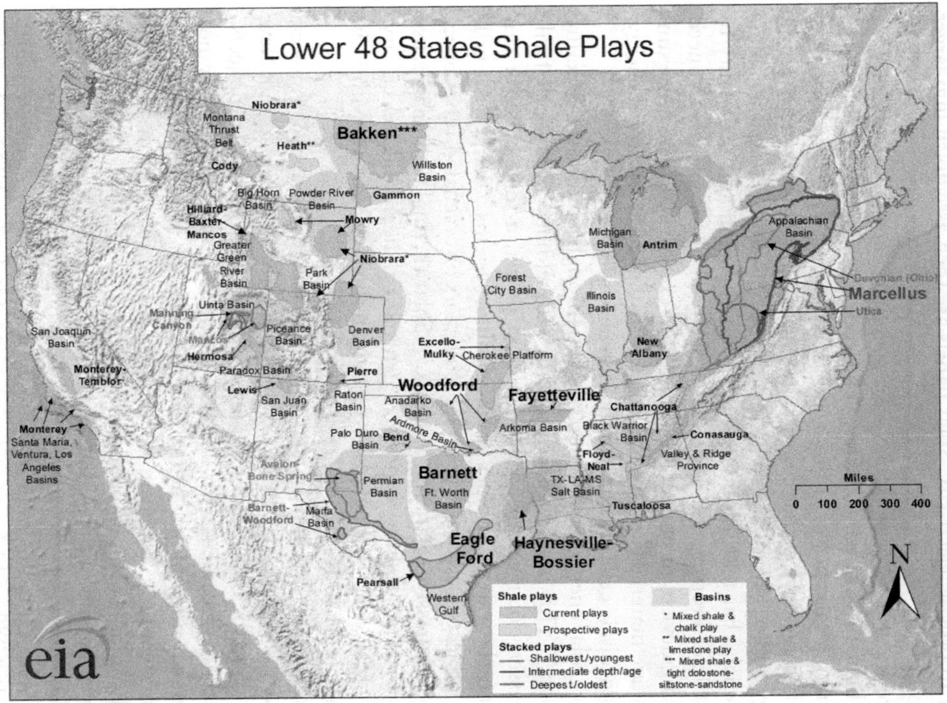

Figure 3.4 American state shale oil and natural gas opportunities.

Source: US Department of Energy, Energy Information Administration, Maps, Shale Gas and Oil Plays Lower 48 States, 2014.

of solar power and many solar power units are connected to an electric grid or an energy storage device in order to address the impact of weather. Solar photovoltaic technology uses semiconductors to produce electricity.[94] When hit by sunlight, a semiconductor material in the form of cells assembled in a panel creates an electrical charge, which is transferred through a circuit to provide electricity. Solar panels are then assembled into larger arrays to produce electricity.[95]

There are four ways to harness solar energy: photovoltaic systems (converting light to electricity), heating and cooling systems (solar thermal), concentrating solar power (utility scale), and lighting.[96] Active solar energy systems employ devices that convert the sun's heat or light to another form of energy and passive solar energy refers to special siting designs or building materials that take advantage of the sun's position and availability to provide direct heating or lighting.[97] Passive solar also considers the need for shading devices to protect buildings from excessive heat from the sun. Photovoltaic materials and devices convert sunlight into electrical energy, and Photovoltaic cells are commonly known as solar cells.

Passive solar energy uses the sun's energy through a building's structure and design.[98] This is a site-specific energy source, not a utility-style facility. If passive solar cannot provide enough power, a solar collector can concentrate heat in water pipes or air handling systems.[99] Large-scale utility solar projects use parabolic trough collectors or evacuated tube collectors to build massive solar array projects.[100] Solar thermal technologies use sunlight to generate heat to power a generator that produces electricity. 12 utility-scale solar thermal plants operate in the United States with 11 in California and 1 in Nevada.[101]

All regions of the United States can enjoy the benefits of solar power. The desert southwest states of California, Nevada, Arizona, and New Mexico enjoy the best location for large scale, utility-grade solar projects. However, urban centers with large concentrations of pollution may not have as much sun and heat as the Desert Southwest but the benefits they gain from solar projects may in fact be larger for the region than solar projects in the middle of an empty desert.[102]

Many successful utility grade solar projects exist across the United States. Boulder City, Nevada's parabolic trough solar array known as Nevada Solar One, was built in 2007 with an annual electricity production of 136 GWh, which means it powers 15,000 homes.[103] Nevada Solar One avoids 129,000 metric tons/year in carbon pollution, covers 130 hectares, and cost $250,000,000 to build.[104] Nevada Solar One is the biggest solar parabolic trough solar plant installed in the last 19 years and the third biggest in the United States.[105] Nevada Solar One's production is 9 percent higher than forecast.[106] The success of Nevada Solar One and other utility-level solar projects are leading major investors such as Warren Buffett to look at solar power as a strong area for investments.[107]

Solar's growth prospects look strong. Utility scale solar projects still are forecast to not even constitute 1 percent of all American utility energy production; utility-scale solar capacity grew by 96 percent in 2013.[108] That growth is expected to continue. Between 2013 and 2015, growth in utility-scale solar projects is expected to be at 52 percent, and customer-sited photovoltaic projects will grow at an even faster rate.[109]

Clear disadvantages to solar power exist. The inconsistent nature of the power source creates major challenges from a reliability and cost standpoint. Outside of the Desert Southwest, the inconsistent nature of solar power calls into question whether the power will be there when it is needed, and dramatically increases the cost. Cost of solar production anywhere is a separate issue of concern. Solar project are typically the most costly form of energy production and happens through regulatory mandate more often than economic necessity.

Wind

Wind is a source of energy that capitalizes on the uneven heating of the atmosphere, irregularities of the earth's surface, and the rotation of the earth.[110] Terrain impacts wind flow patterns and different regions of the United States enjoy different wind flow patterns and speeds.[111] Wind, like solar, is an intermittent source of power as wind may not be strong enough all the time to produce power.

Wind develops energy through the use of wind turbines that converts kinetic energy into mechanical power.[112] Wind turbines are either horizontal-axis or the vertical-axis design.[113] Horizontal-axis wind turbines operate with two or three blades that must face the wind. Wind turbines operate on land or offshore in large bodies of water.[114] Utility-scale wind turbines are as small as 100 kW to as big as several MWs and, like solar arrays, are more efficient when grouped together.[115]

Although all wind turbines operate on similar principles, several varieties are in use today. These include horizontal axis turbines and vertical axis turbines, but horizontal axis turbines are the most common turbine configuration used today.[116] They consist of a tall tower, atop which sits a fan-like rotor that faces into or away from the wind, a generator, a controller, and other components, and are built with two or three blades.[117] Horizontal axis turbines sit high atop towers in order to take advantage of the stronger and less turbulent wind at 100 feet (30 meters) or more aboveground, and each blade acts like an airplane wing; so when wind blows, a pocket of low-pressure air forms on the downwind side of the blade.[118] The low-pressure air pocket then pulls the blade toward it, which causes the rotor to turn—this is called lift.[119] The force of the lift is actually much stronger than the wind's force against the front side of the blade, which is called drag, the combination of lift and drag causes the rotor to spin like a propeller, and the turning shaft spins a generator to create electricity.[120]

Wind energy is also benefiting from technological innovations. Wind turbines are becoming bigger, longer, and operating with lighter blades and improved performance. The average wind turbine capacity increased by 170 percent since 1998, and the price for wind production declined to a highly competitive 4 cents/kWh.[121]

The US wind industry broke all previous records by installing nearly 10,000 MW of new generating capacity in 2009 (enough to serve over 2,400,000 homes).[122] These new projects place wind power neck and neck with natural gas as the leading source of new electricity generation for the country.[123] Together, the two sources account for about 80 percent of the new capacity added in the year; however, 2010 saw a decline in new wind power capacity in the United States as the economy slowed.[124] In 2012, wind energy was 43 percent of all new electricity additions with $25,000,000 in US investments, and wind became the top source of new US electricity generation capacity.[125] 2012 saw the addition of over 13 GW of new wind power, constituting a doubling of wind capacity from 2011.[126] Wind's 60 GW of power generation is enough to power 15,000,000 homes—that is, all the homes in California and Washington.[127] That is a 22-fold increase since 2000.[128] The wind industry is also a job creator. 72 percent of the wind turbine equipment installed in the United States was made domestically.[129] This number was 25 percent in 2006–2007.[130]

Of course, not all regions in the United States have equal access to wind as a source of power (figure 3.5).

Texas uses more wind energy than any other state and they have twice the amount of number second-placed California. The Rocky Mountain and Great Plains regions have the greatest natural wind development opportunity. Nine states rely on wind power for more than 12 percent of their electricity, and Iowa, South Dakota, and Kansas use wind for more than 20 percent of their electricity.

Even the renewable wind energy industry faces aggressive opposition. Environmental concerns about the impact on birds, noise, and visual challenges, and questions about the inconsistent and inefficient delivery of electricity exist. Offshore projects offer high wind speeds but even higher costs, and have not taken off yet.

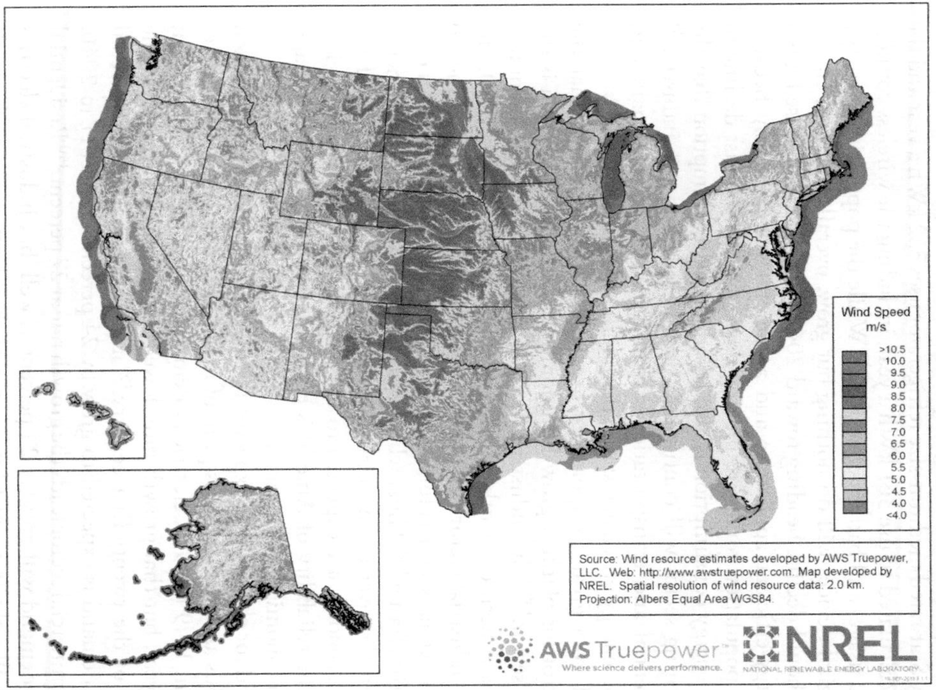

Figure 3.5 United States land and offshore based annual average wind speeds.

Source: This image website has been reprinted with permission from the National Renewable Energy Laboratory (http://www.nrel.gov/gis/images/100m_wind/awstwspd100onoff3-1.jpg). Accessed December 18, 2014.

Demand for Energy

Do you think a refrigerator or new iPhone uses more electricity? It is the iPhone.[131] In fact, the digital revolution that American consumers and businesses are enjoying is now using 10 percent of the energy in the nation.[132] An Energy Star rated medium-size refrigerator uses 322 kWh a year and the iPhone is calculated to use 361 kWh a year, including the wireless connections, data usage, and battery charging.[133] While the popularity of energy driven digital devices will do nothing but grow, overall American energy demand should decline heading toward 2040. This decline is based upon more energy-efficient appliances and automobiles, but mainly because America's population totals are moving down.[134] The forecast decline in American energy consumption is also based upon an assumption that the manufacturing sector will continue to decline as a domestic industry.[135] The residential and transportation sectors are expected to remain flat, based upon energy efficiency steps taken in these markets.[136]

It is of even greater importance to the American economy that the largest sector for planned growth in domestic energy consumption is the industrial sector.[137] The industrial sector houses the high-wage manufacturing jobs America is working hard to keep. From 2011 to 2040, the American industrial sector will grow by 5.1 quadrillion BTUs, and the job heavy commercial sector is forecast to grow by 3.1 quadrillion BTUs over the same timeframe.[138] Thus, while overall American demand for energy will remain flat over the next 30 years, the high-wage job sectors of the American economy are expected to illustrate some levels of growth. The cost and the reliability of America's energy demand will remain a major factor for economic development.

Demand for electricity is expected to continue, but not at the rates seen in the 1950s as America hit its peak of industrial production and job creation. From 1949 to 1959, electricity demand grew at almost a 10 percent rate each year, but that growth rate slowed to just under 1 percent in the early part of the twenty-first century.[139] However, total American retail electricity demand is expected to grow by 24 percent from 2011 to 2040, and the commercial sector leads the growth rate at 27 percent.[140] Residential electricity demand will grow at 24 percent as well, but industrial electricity usage will only grow by 17 percent over this thirty-year timeframe.[141] High economic growth areas will experience greater demand for electricity, while low growth regions will experience less electricity usage.[142]

Demand for natural gas is expected to rise. The global demand for natural gas will increase by 50 percent before 2035.[143] This demand is driven by the massive new supply of natural gas created by technological innovations to permit additional oil and natural gas to be gained from shale deposits across the world. Shale gas is also expected to influence the growth prospects of coal as an energy source. Shale and federal government environmental regulations are impacting the use of coal in the United States and Europe. However, emerging economies such as China, India, and others will expand their coal usage by one-third, but coal's global demand as an energy source will move from 41 percent now to 33 percent by 2035.[144]

Demand for oil is again being driven by China and emerging economies. As the Chinese middle class grows, so too grows their purchasing power. Much like the United States 100 years ago, the Chinese consumer market is expanding rapidly into being able to afford modern appliances, wireless phones, computers, and the almighty automobile. All these devices need energy, and the automobile of course is powered by oil. China is forecast to overtake the United States as the world's largest consumer of oil by 2030, and the Middle East will also experience growth in its demand for oil.[145]

The global energy demand is not going down but is headed up, and is slated to increase by one-third from 2011 to 2035.[146] Emerging economies in Asia and the Middle East are seen as major drivers for this increased energy demand. India will join China as a major Asian center for energy demand.[147] The Middle East natural gas market is expected to grow more than will all of Europe combined.[148] The Middle East should be the second largest natural gas consumer by 2020 and, by 2030, the third largest oil consumer in the world.[149]

Another illustration of global demand opportunities is the 1,300,000,000 people worldwide who currently lack access to electricity and the 2,600,000,000 people who use biomass for cooking.[150] 95 percent of this population is located in Asia and sub-Saharan Africa.[151]

Nevada and Michigan's Energy Led Economic Development Strategy

Nevada and Michigan are searching for economic answers through energy. Since 1983, the state of Nevada has been searching for ways to diversify its economy.[152] A recognizable brand name, in a warm weather climate, in a Right to Work state with no state income tax in a growing region makes Las Vegas an attractive market for private sector investment. Nevada launched an energy led economic development strategy with a range of programs illustrated in table 2.1 (figure 3.6).

Nevada offers 25–35 percent tax abatements for 5–10 years, depending upon the level of energy conservation achieved by the green building.[153]

- Building Energy Codes
- Commercial Retrofit Grant
- Energy Planning & Conservation Fund (NDOW)
- Nevada Portfolio Energy Standards
- Green Building (LEED) Tax Abatements
- Public Facilities Retrofit Grant
- Renewable Energy Tax Abatements
- Revolving Loans for Renewable Energy, Energy Efficiency, and Energy Conservation
- State Energy Program Formula Grant
- UNLV Desert Research Institute

Figure 3.6 Nevada's energy programs.

Source: Database for State Incentives for Renewables and Efficiency, retrieved from http://dsireusa.org/incentives/index.cfm?re=0&ee=0&spv=0&st=0&srp=1&state=NV.

Larger—55 percent—tax abatements for 20 years are available for new or expanding Nevada companies who generate electricity from renewable sources such as solar, wind, biomass, fuel cells, geothermal, or hydro with a generation capacity of at least 10 MW.[154] These companies must create a minimum number of jobs and make a substantial capital investment for only private sector business projects.[155] In addition, Nevada offers a sales tax abatement for the purchase of renewable energy equipment used to generate electricity.[156] Again, the generation facility involved must be 10 MW in size and the project must involve job creation and capital investment.[157]

Nevada uses federal money to provide short-term, low-cost loans to developers of renewable energy projects, renewable component manufacturers, and energy efficiency and conservation projects.[158] Nevada awards small grants to support the development of renewable energy projects at schools and other public institutions.[159] PACE financing, permitting property owners to borrow money to pay for energy improvements based upon the savings from the improvements, is also permitted in Nevada.[160] Nevada permits local governments to use PACE with payment structured through special assessment-charged property owners.[161]

Nevada is pushing its energy companies to embrace renewable energy through a regulatory portfolio energy standard. Nevada requires their major incumbent electric utility provider, NV Energy, to derive or save a percentage of the electricity it sells from renewable energy resources or energy efficiency measures.[162] Las Vegas works on enhancing their energy led economic development approach through research and development. UNLV's Desert Research Institute is a leading R&D center for clean energy, energy efficiency, and water conservation. The Desert Research Institute gained nearly $400,000,000 in research and development funding since 2000 and provides a support element for renewable energy companies in the state.[163]

Michigan and Energy Led Economic Development

Michigan is not ignoring energy led economic development either. Michigan's Governor Rick Snyder laid out a "no regrets" energy policy centered on four pillars, including reducing the state's reliance on coal as an energy source and moving to natural gas and renewable energy sources, improving the reliability of the state's energy supply, reducing the energy costs for the state's industrial base, and reducing the environmental impact of energy production and use.[164] On the energy efficiency front, Michigan Saves is a partnership supported with public funds but working with private lenders to provide low-cost financing to help residents and businesses save energy and money. Since late in 2010, more than 1,700 Michiganders have used Michigan Saves funds to reduce energy costs by about $350 a year on average.[165] Michigan is also looking to create a Strategic Natural Gas Reserve for Michigan to capitalize on state owned natural gas deposits.[166]

Michigan uses a wide range of renewable energy mandates and incentives in an effort to diversify its energy sources, promote cleaner energy sources, as well as recruit renewable energy companies. Michigan's Energy Efficiency and Renewable Energy Revolving Loan Fund Program supports renewable energy and energy efficiency investments by small manufacturing companies by providing them with low-interest government loans.[167] Alternative energy research and development or manufacturing may use a nonrefundable tax credit from the Michigan business tax.[168] This tax credit is larger if used within one state's economically distressed regions—known as Renaissance Zones. Tax Credit eligible technologies include fuel cells, PV, biomass, solar thermal heating and cooling, wind energy, CHP, microturbines, miniturbines, Stirling engines, electricity storage systems, and clean fuel energy systems powered by methane, natural gas, methanol, ethanol, or hydrogen.[169] Also, companies locating in designated Next Energy Zones and performing research and development or manufacturing of alternative energy technologies may claim a refundable payroll tax credit.[170]

Modeled after the national concept of an enterprise tax abatement zone, Michigan's Renewable Energy Renaissance Zones offer tax benefits such as not paying the state business tax, state education tax, personal and real property tax, or local income tax for up to 15 years, with abatements being phased out in 25 percent increments over the last three years of the zone designation.[171] Michigan looks to recruit renewable energy facilities that create energy, fuels, or chemicals directly from the wind, the sun, trees, grasses, biosolids, algae, agricultural commodities, processed products from agricultural commodities, or residues from agricultural processes, wood or forest processes, food production and processing, or the paper products industry, solid biomass, animal wastes, or landfill gases.[172] These facilities can focus on research and development as well as manufacturing.[173] 15 Renewable Energy Renaissance Zones are permitted in Michigan.[174] Detroit offers loans between $50,000 and $150,000 to commercial, institutional, and public organizations to install energy efficiency and renewable energy technologies.[175] Detroit's program is funded with federal dollars, in partnership with the city's Economic Development Corporation.[176] Like most states, Michigan also offers PACE financing and permits local governments to provide PACE financing loan programs for renewable energy and energy efficiency investments.[177] In a related program, Lean and Green Michigan supports industrial, commercial, and multi-family property owners gain energy audits to develop cost effective energy efficiency strategies.[178] Five counties and three cities have participated in the program.[179] Consumers Energy offers an Experimental Advanced Renewable Energy Program offering customers a buy-back tariff program for electricity produced by solar photovoltaic systems and anaerobic digestion.[180] Michigan, joining Nevada and several other states, also joined the Right to Work Movement, and now has a law that permits private sector employees not to join a union even if their workplace votes to do so. This matters for some manufacturers—especially outside of the union-heavy auto industry.

Michigan has a state renewables portfolio standard, a net-metering law, and a Wind Energy Resource Zone Board to promote the use of renewable energy and energy efficiency.[181] Michigan set a goal of reducing state government grid-based energy purchases by 25 percent before 2015, based upon a 2002 baseline; thus, the state launched a successful energy efficiency program for its own facilities.[182] Michigan's electricity suppliers are required to disclose to customers what their fuel mixes and emissions are for the energy they use.[183] Furthermore, Michigan has renewable portfolio standards and a mandatory, statewide net-metering program. Michigan's net-metering law applies to traditionally rate-regulated utilities (investor-owned and rural electricity cooperative) and alternative electric suppliers.[184] A wide range of renewable energy sources are eligible for net metering.[185] Finally, Michigan requires the state's investor-owned utilities, alternative retail suppliers, electricity cooperatives, and municipal electric utilities to generate 10 percent of their retail electricity sales from renewable energy resources by 2015.[186]

Big Questions

1. Do all regions of the United States have a unique supply of energy?
2. Should regions and states focus on renewable and nonrenewable sources of energy?
3. Does any form of energy operate without a government subsidy?
4. How do policymakers justify promoting high-cost forms of energy?
5. Is shale gas leading an American energy revolution or delaying the transformation to renewable energy sources?

Class Debate

The class will continue their exercise to support the creation of a strategic energy led economic development plan for Michigan or Nevada. Class discussion will focus on how energy led economic development can address these two states' economic futures and on a discussion of the energy led economic development strategies and tactics they should adopt.

Class Writing Assignment

Draft five strategies and ten tactics for the Michigan or Nevada strategic energy led economic development plan.

Additional Readings

2012 Wind Technologies Market Report, US Department of Energy, August 6, 2013.

Annual Energy Outlook 2014, US Department of Energy.

Bullis, Kevin, "Can Small Reactors Ignite a Nuclear Renaissance?" MIT Technology Review, March 28, 2013.

Galbraith, Kate, "Retrofitting Dams to Generate Electricity," *New York Times*, August 19, 2009.

"Geothermal Heat Pumps in K-12 Schools: A Case Study of the Lincoln, Nebraska Schools," Oak Ridge National Laboratory, ORNL/TM-2000/80, February 4,

2013, p. 12, retrieved from http://www.ornl.gov/sci/ees/etsd/btric/pdfs/com_ghpsinschools.pdf.
Hadjerious, Boualem, "An Assessment of Energy Potential at Non-Powered in the United States," US Department of Energy, April 2012.
"Maine Project Takes Historic Step Forward in U.S. Tidal Energy Deployment," US Department of Energy, News Report, May 4, 2012, retrieved from http://energy.gov/articles/maine-project-takes-historic-step-forward-us-tidal-energy-deployment.
Roenker, Jonathan M., "The Economic Impact of Coal in Appalachian Kentucky," Center for Business and Economic Research, University of Kentucky, 2001.
Siler-Evans, Kyle, Inês Lima Azevedo, M. Granger Morgan, and Jay Apt, "Regional Variations in the Health, Environmental, and Climate Benefits of Wind and Solar Generation," *PNAS 2013*, June 24, 2013.
"Top 10 Things You Didn't Know about Hydropower," US Department of Energy, Energy Information Agency, April 19, 2013, retrieved from http://energy.gov/articles/top-10-things-you-didnt-know-about-hydropower.
Tverberg, Gail E., "Oil Supply Limits and the Continuing Financial Crisis," *Energy*, 37, 1 (January 2012): 27–34.
US Energy Department, Energy Information Agency, Coal Report, 2014.
What Is Shale Gas and Why Is It Important? US Department of Energy, Energy Information Agency, Energy in Brief, December 5, 2012.

Chapter 4

Energy Led Economic Development Case Study

Chapter Goals

1. Understand the role of the case study in the learning process.
2. Understand how public policy is made.
3. Take on the roles of policy researcher, advocate, and negotiator in the economic development context.
4. Recognize the players in the Energy Zone Case Study.
5. Understand all sides of the argument.

The Case Study Process

Case studies illustrate how the real world operates and requires research of legal aspects, policy and business issues, strategy development, policy proposal creation, and advocacy for those positions. Policymakers decide upon a region and state's direction through the case study in this text, and their work began with the development of a strategic energy led economic development plan for Michigan and Nevada. The case study runs through the implementation of that plan to create an urban energy center, rural energy center, address land use policies, and address the cost of their energy, all with the goal of energy company retention and attraction campaigns.

Public Policy Creation and Energy Led Economic Development

The public policy formation process impacts the development of energy led economic development policies and involves six distinct steps geared toward solving a societal problem.[1] The first stage of public policymaking is the identification of problems and opportunities, such as high unemployment, low wages, air quality, and energy costs.[2] Regions and states have opportunities that public policy actions can better position their community to succeed.

Once a problem or opportunity is identified but before solutions are determined, the issue competes for priority on the region and state's public policy agenda.[3] This agenda-setting exercise is impacted by interest groups as well as government officials and energy led economic development issues compete with all other policy issues before the government. What complicates energy led economic development policy formation is the fact that many of the stakeholders in the energy industry have polar opposite views about which energy sources communities should utilize. Next, policies are created to address the challenges identified. The public policy process involves legislative enactments of city councils, county officials, state legislatures, and the Congress, as well as interaction with the executive branch and, potentially, the judiciary. Public policy formation is impacted by budgeting and fiscal efforts.[4] Executive agency officials submit a budget on an annual or biannual basis that funds many of the public policy initiatives. These budgets generally are introduced as legislation and debated by legislative bodies. Policy implementation can include rule adoption, regulations, service provisions, education, and mediate disputes. Finally, public policy strategies are evaluated for their effectiveness. Executive and legislative bodies may be involved in public policy program evaluation. Legislative bodies provide oversight of existing programs through hearings, and executive agency departments provide policy analysis to determine if the policy solution is working.[5]

Role of Public Relations

The news media, whether newspaper, television, radio, or the Internet, plays a substantial role in developing energy led economic development policy. Framing theory promotes the concept that how an issue is characterized influences how it is understood.[6] Quite often, the substance of the news story is not as important as the decision to cover events from a particular perspective. What events news editors decide to cover, what kind of placement the story is given, who the reporter interviews, and who they do not all frame a story and shape its impact on energy led economic development. The news media not only covers issues, but also provides editorial content to provide their readers with opinions on relevant issues. Energy led economic development policies are impacted by the news media every day, with news coverage ranging from land use developments to coverage and editorial debate over the legislative process.

Role of Service Firms

Professional service firms impact energy led economic development strategies, as they are often the companies asked to turn a vision into a reality. Services firms involved in energy led economic development include design professionals and lawyers. Design professionals and land planners play a critical role in the land use approach for an energy led economic development initiative. These professionals define a developer's vision. These professionals are part businessperson, part artist, and they provide the visual that image developers need to turn their vision into a reality. Design professional and land planners in the energy led economic development strategy context can design sustainable developments, promote mass transit, find a place for the almighty automobile, and craft a plan office, retail, industrial, and residential developers want to locate in.

Lawyers also are central to energy led economic development. Lawyers' understanding of the lawmaking, regulatory and judicial process puts them in a unique position to negotiate energy led economic development issues. This can range from negotiating energy company incentive contracts to utility regulatory structures to land use deals.

The Interest Groups

A wide range of public and private sector interest groups impact energy led economic development strategies. Business leadership is engaged in the energy public policy debate. Business leaders can have a financial basis for being involved or playing the role of the "honest broker." Business leaders may be on all sides of the energy issue. They may be part of the "fossil fuel crowd" promoting traditional natural gas, oil, and coal as the low-cost energy alternative. They may be part of the "renewable energy" crowd promoting wind, solar, geothermal, biomass, and other renewables as the key to economic success. Finally, they may represent the great majority of business owners who don't produce energy but use it.

Developers in the context of energy led economic development are for-profit companies or not-for-profit government organizations focused on improving the use of land for their economic gain. Developers play a vital role in energy led economic development by bringing their private sector real estate development know-how to understanding what the private sector market will find attractive when it comes to land use investment.

Environmentalists are the heart and soul of an energy led economic development movement. They may be the biggest advocates or the loudest opponents of an energy led economic development strategy. Environmentalists are driven by an ideological belief that the Earth has limited resources and it is the job of humans to protect those resources.

Unions are a stakeholder in energy led economic development strategies. While the percentage of unionized workforce is on the decline, many regions and states are home to labor unions. Existing or retired union membership is a large group of voters many policymakers are interested in supporting. Labor unions impact economic development projects in many cities across the United States, as they may be interested in having their members build the project or work in the planned facility.

Neighbors can be for or against an energy led economic development project in their neighborhood. Opponents to public policy are usually louder than proponents. Not In My Backyard—NIMBY—is an acronym that epitomizes the perspectives of opponents to public policy issues. Preservationists are engaged in energy led economic development policy formation. Advocates for historic preservationist want old buildings restored to their previous glory and older neighborhoods redeveloped. Cost does not matter to them because they usually do not own the buildings. The Tea Party is a new force in energy led economic development policy formation. They are fiscally conservative but do not trust large companies. They impact the energy led economic development debate and often will be focused on the cost to taxpayers of the project at hand.

Role of the Corporate Site Location Process

Energy led economic development projects all hope to have numerous companies grow or move to their region because of the initiative. This economic growth occurs through a corporate site location process, where companies negotiate an incentive package. This multistep process starts with a company defining the business case for where it could potentially locate. The company examines its current business opportunities based upon its facility locations, employee head count, capital investment, community connections, and taxes paid. If a business case exists, the company then determines their potential job growth and capital investment, generally over a three-year timeframe in the near future as well as the actual plans for the facility. Facility plans address square footage of the building, operational requirements, capital cost budget, parking, power, water, sewer, rail, road, and other infrastructure needed at the site. Based upon supply chain, customer growth, regulator, tax, and workforce issues, the company identifies potential regions, states, or nations that they wish to consider for a potential facility location.

Following all this research, a consultant will send a letter on behalf of the company, not identifying themselves but outlining the potential of an economic development project to local and state economic development leaders. This letter defines the space and location needs of the project, and makes specific incentive requests for the project. Local and state economic development officials are charged with answering these corporate site location letters with specific site proposals and economic development incentive offers. The site selection consultant and company negotiate an economic development incentive package to gain tax incentives at the site that best fits their business case. Finally, the company participates in a public announcement with government leaders announcing the deal and the local and state government officials approve or disapprove the deal following its announcement, after a public hearing conducted by a legislative or executive agency body.

Global Energy Center (GEC) Case Study: GEC 2020

The GEC Case Study is set in Nevada and Michigan and involves local and state economic development leaders redeveloping this region and state around an energy led economic development strategy. GEC is focused of these two states because they have different economic challenges and opportunities but both are looking to energy led economic development for answers to those challenges and opportunities. The GEC hopes to capitalize on multiple sources of energy discovered in the region to build a diversified economy in a range of industry sectors. GEC involves multiple cities and states, who at times compete with one another and at other times work in partnership. GEC will involve the development of a larger energy led economic development strategy, business decisions about what industries to focus on, and multiple public policy debates on zoning, annexation, eminent domain, tax policy, infrastructure finance, workforce development, and quality of life issues. All of these policy areas impact the development of a long-term, sustainable energy led economic development strategy that will outlast the region's supply of energy.

GEC is the brainstorm of the business leadership from Nevada and Michigan created in an effort to capitalize on the recent discovery of shale gas and crude oil and renewable energy sources in the region and to position these states and their big cities and small towns for growth based upon this energy production and beyond. GEC proposes to develop a multi-billion dollar energy led economic development strategy that capitalizes not just on the energy in the region but takes advantage of the natural physical infrastructure of the Great Lakes and access to the Western US market. GEC is just an idea in need of a strategic plan and positive decisions by policymakers in these states to achieve success. GEC will also have to compete with others around this idea.

GEC's Energy Led Economic Development Opportunity for the Nevada and Michigan

Discovery of shale oil and natural gas deposits and opportunities with renewable energy sources create an opportunity to use an energy led economic development strategy to make Michigan the global economic leader it once was. While Michigan lacks these shale oil deposits, its neighbor Ohio has them, and Michigan is working out how to capitalize on them. Ohio's natural gas and crude oil industry reinvested approximately $238,000,000 on new energy exploration and development in the state, based primarily upon the shale play in the state that is driving domestic oil and natural gas discoveries all across the United States.[7] This is making the Ohio oil and natural gas industry a billion dollar industry in the Buckeye state, which has distributed more $1,000,000,000 in royalty checks to Ohioans since 2000 for the 275,000 wells that have been drilled. The Ohio oil and natural gas industry is paying over $30,000,000 in local, state, and federal taxes annually, supporting almost 5,000 direct jobs and nearly 13,000 indirect jobs constituting nearly a billion dollars in salaries and income in Ohio alone. Ohio consumers are also benefiting from $30,000,000 in savings from lower natural gas and crude oil costs. The Ohio shale boom is replicated in western Pennsylvania and is producing an energy led economic development opportunity. Michigan is watching this Ohio boom and looking for ways to capitalize on it themselves. Michigan's strategic location on the Great Lakes—connected directly to global markets through Canada—may offer them a logistics and distribution opportunity that could serve as a launching pad for companies searching for cheap energy.

Michigan is also connected to a strong renewable energy market in the Midwest. Wind energy and biomass are leading Midwest energy sources. More than one-third of America's wind energy capacity if located in the Midwest and, of the nine states that produce at least 10 percent of their energy from wind, five are in the Midwest.[8] Wind capacity in the Midwest grew by more than 29 percent in 2012.[9] 80 percent of the nation's biomass is produced in the Midwest as it is part of the nation's farm belt.[10] The US Department of Defense is funding a jet fuel biofuels research project in Illinois, and similar efforts have occurred at the Air Force Research Lab in Dayton's Wright Patterson Air Force Base.[11] While the production of domestic solar panels has been hit hard by Chinese competition, Toledo, Ohio remains a global leader in solar research and development. Major research and development efforts around renewable energy at the University of Michigan, Ohio State University, and others are developing cutting-edge applications for the renewable energy industry and working to tie twenty-first century energy sources to nineteenth century inventions such as the automobile. Michigan is leading the charge on fuel cells and other Midwest states are working to be positioned as global renewable energy leaders.

Nevada and its largest city Las Vegas aspire to become known as Solar City. The heat—and days of intense sunlight—makes southern Nevada the perfect place for the use of solar energy. In addition, the use of electricity in Las Vegas is massive. Casinos and homes all run on electricity and they are looking for new sources and cheaper sources of energy. Las Vegas, like Detroit, is hoping to diversify its economy with industries beyond gaming and government, which make up the vast majority of the region's employers. Las Vegas and Nevada have many pro-growth policies already in place in that they lack an income tax and are a Right to Work state—both of these policy decisions are attractive to many potential employers. The Las Vegas airport, like Detroit's airport, connects customers to the world with direct flights. This airport infrastructure is critical to advanced service companies looking for national and global connections being just a direct plane flight away. Las Vegas is a blank slate for economic development and their conservative state and local government create a friendly atmosphere for economic growth.

However, both Detroit and Las Vegas have costly urban sprawl and a struggling school system to deal with. While Michigan's system of higher education is strong, Nevada's needs some work. Nevada is behind in the production of college graduates in whom advanced service companies are interested. However, no one can compare the economic challenges of Las Vegas with those of Detroit.

GEC Players

The GEC players include local and state government officials, the region's business leadership, environmental activists promoting renewable energy, oil, and natural gas companies promoting fossil fuels, Tea Party activists watching government spending, and Smart Growth advocates focused on reducing the urban sprawl of the region. Many agendas are pushed and debated upon, all with the goal of improving the region's economy.

Governors are the highest-ranking elected officials, not just in state government but also in the state as a whole. Maybe a high profile US Senator dominates the political scene for a while, but typically, it is the Governor that drives public policy and works with the business community to address the challenges to the state and region while capitalizing on its opportunities. The Governor works not just with business leaders but also with the local government leaders, from major urban centers and small, rural communities. Ohio, Pennsylvania, and Michigan in the GEC Case Study have three different Governors, and getting them on the same page may prove a challenge. Michigan's Governor is straight out of the labor union. Michigan is a political swing state that can be dominated by either party, but organized labor is at its core. The United Auto Workers (UAW) still holds much sway, and, after the election of a couple Republican Governors, they finally got their man back in the Governor's mansion with the election of a Congressman from Detroit with strong ties to organized labor. Michigan's Governor Walter Reuther, IV is a great, great grandson of the founder of the UAW, but he was still elected with the mandate to do something about the declining Michigan economy. The Governor of Nevada is a Chamber of Commerce Republican elected to address the states' struggling economy. He supported Nevada's renewable energy policies. Nevada Governor John D. Rockefeller, IV is a great, great, great grandson of the oil magnet who used to call Cleveland home. Rockefeller, who won the Governor's race "the old fashion way"—he bought the Governor's race with $25,000,000 of his own money—realizes energy is the answer. Michigan and Nevada have Republican dominated Statehouses. Each state's major cities are in economic distress, but have opportunities in terms of energy growth.

Various interest groups are engaged in the GEC Case Study. The GEC Committee is a group of national business leaders looking to promote energy led economic development in a couple of targeted states before rolling out the approach nationwide. These business leaders are civic-minded executives with a focus on the major business and policy issues impacting their markets. Business leaders, on the good side, put money and resources behind economic development efforts they support but, on the bad side, sometimes view economic development as completing major real estate projects and not building a sustainable economy. Business leaders are also interested in benefiting from lower energy prices through the capturing of new, low-cost energy sources or regulatory changes influencing the cost of electricity and natural gas that power their facilities.

The National Taxpayers League (NTL) is composed of conservative residents generally opposed to government spending, and they question when large institutions try to design economies. They challenge the business leaders and elected officials because they are rarely "for" anything, but they can be popular with the public as they get media attention. NTL sees their role as challenging public officials and not accepting as gospel what the business community dictates. GEC will not happen without the help of real estate developers. GEC needs viable economic development projects that provide private sector jobs to achieve success, and real estate developers can make those projects happen. Real estate developers compete with each other and will do so as part of GEC. They do not like government but are professionals at gaining funding from local, state, and federal government sources. Neighbors surrounding potential developments are another major player impacting GEC. Supportive neighbors are not the ones that show up to discuss economic development projects—unhappy people travel and speak up in public meetings. Not In My Backyard (NIMBY) is the term used for neighbors opposed to a development who just might want to benefit from its services but do not want any of the traffic or other negatives from the project.

Big Labor is a relevant player in the GEC. The percentage of Americans in labor unions is low, but their political clout still matters. Unions have current and retired members well placed in Nevada and Michigan based upon these states' roles in the gaming and manufacturing industry respectively. Big Labor could be an advocate for an energy led economic development strategy, but that likely depends on whether their members would have jobs in the project at hand. New Urbanists' Unite (NUU) is an organization devoted to the creation of compact land use patterns and the redevelopment of the urban core. This group of land use planners, citizens, and "good government" types are supporters of Smart Growth and see this strategy as an essential part of any energy led economic development strategy. From their perspective, how can an energy led economic development strategy be successful if it wastes energy in its very design?

Energy companies are of course major players in the GEC. Traditional energy utilities are central to the policy debate around energy efficiency and which source of energy policymakers wish them to use. Oil and natural gas companies are focused on promoting the use of affordable fossil fuels that have recently become available in Ohio and Pennsylvania. Renewable energy companies are focused on the wind, biomass, and solar opportunities the region possesses. In theory, GEC should include all these sources of energy but these energy companies do not play well together.

GEC Facts

The GEC Committee decided that the future of the American economy lies in energy led economic development. Rather than go to Washington and battle the President and the Congress, GEC Committee decided to build energy led economic development strategies at the state and regional levels first. Nothing sells better than a couple of successful regional and state economic development models. The goal is to make two states and their major cities a shining example of how an energy led economic development strategy can transform an economy. The strategy centers not just on attracting energy companies, but also in further diversification of the regional and state economy. Furthermore, GEC's goal is to soften the boom and bust of energy productions in rural communities and develop two global energy centers in urban communities. GEC also is interested in promoting all forms of energy.

The private sector is excited about the GEC and is jumping into the project with both feet. GEC supports a Wind Michigan proposal to develop the world's largest windmill farm in the Lake Michigan area, connected to a small rural community in the hopes of developing a wind industry cluster of companies. Wind Michigan's proposal is popular with the alternative energy industry and, to be accomplished, must overcome land use and technical issues; but Wind Michigan wants Michigan to adopt a long list of regulatory and tax incentives. These regulatory and tax incentives cost money and are not popular with traditional incumbent utilities nor the heavy manufacturers that question wind's ability to produce the consistent power they need. Michigan is also focused on its rural communities. The Michigan Ag Association's strategy involves a new tax on utilities to pay for a massive batch of biomass tax incentives to fund the largest development of utility scale, biomass power facilities in the United States. The utility industry is unexcited about the proposal but the local communities and biomass industry are supportive. Even more supportive is the Governor who is working to not let the Buckeye state's shale play make the Wolverine state look bad.

GEC supports the proposal by a Texas pipeline company to build the Wolverine Pipeline to take shale gas discovered in Ohio and ship it to a facility in rural Michigan. The Wolverine Pipeline is key to the state's recruitment of energy-intensive companies hoping low-cost energy will serve as a prime recruitment tool. Prime 101 is a chemical manufacturing searching for a location near cheap energy. Wolverine Pipeline received approval from the federal government for its construction; however, members of the Ohio General Assembly are not happy Michigan is gaining on the shale play. Prime 101 will employ 200 people but it uses a substantial amount of energy. The Detroit Electric Company has another idea for Prime 101—to run on electricity and not natural gas. Detroit Electric Company proposes to use special infrastructure funding to recruit Prime 101, and the City of Detroit and state of Michigan have an array of energy and other economic development incentives to push through at Prime 101.

Finally, GEC supports the development of a Detroit Energy Center with the goal of creating a self-sufficient energy district and other incentives to serve as a prime location for energy-related manufacturers and service companies. The Detroit Energy Center is proposed in partnership with the University of Michigan on 2,000 acres of formerly tax-delinquent land to build a world-class energy-based research and development facility. The site will be energy self-sufficient and could be the spark Detroit needs to succeed.

GEC is working with Nevada on an energy led economic development strategy as well. Las Vegas is trying to become a global manufacturing center. To achieve that goal, GEC supports a Manufacture Vegas initiative to reduce overall energy costs focused on energy efficiency and regulatory changes to how its electric and natural gas industry is regulated. The utilities are against the effort as the energy efficiency proposal provides for utilities to rebate customers back millions of dollars for energy efficiency investments. GEC is also pushing Nevada to address its workforce issue. GEC proposes to create an energy workforce strategy in partnership with the energy industry. Addressing the region's lack of college-educated graduates is a major challenge for the state's effort to recruit advanced manufacturing, services, and technology jobs in or out of the energy industry. The Nevada energy workforce proposal will create an industry-based training program with $50,000,000 appropriated by the Nevada legislature. Much debate is expected. The Tea Party is advocating against the use of taxpayer dollars for this purpose while energy companies are supportive. GEC is focused on addressing the cost of urban sprawl in Las Vegas. At GEC's urging, the Mayor of Las Vegas has proposed that a "smart growth initiative" be instituted in not just Las Vegas, but also in the surrounding counties, in an effort to force growth and development back into the city and to prevent the explosion of urban sprawl in southern Nevada. Smart Vegas is the Mayor's proposal and is modeled after Portland, Oregon's successful development ring program that drives growth and investment in and not out. Needless to say, the development community is unexcited about the Mayor's smart growth proposal but the existing suburban communities see Smart Las Vegas as an opportunity not just to support the development of Las Vegas and attract millennial workers to the region.

Finally, GEC does not believe addressing energy efficiency will be enough to drive down energy costs. Thus, it is supporting the Las Vegas Mayor's proposal to boost the struggling solar power industry and reduce the region's cost of energy by creating a municipal electricity company and developing solar farms for energy. Marketing southern Nevada as Solar City is only part of the effort. The Mayor wants to directly support the solar industry, and is proposing the formation of a public power entity as part of the city government who will build utility scale solar power facilities in which city residents and businesses can buy their electricity services. Solar City Power is a controversial proposal as it looks to insert government into the businesses that have been private, investor-owned utilities for decades. Solar City Power is modeled after the TVA, but about sunlight instead of water. Solar City Power will boost the solar industry through massive direct purchases and provide electricity to residents and businesses without worrying about a profit going to shareholders. Public power is active in many places across the nation, and the Mayor of Las Vegas is making his play in this arena.

Big Questions

1. How can a case study help people understand how to grow an energy led economic development strategy?
2. Can Governors lead the private sector or is it the other way around?
3. How will the three states interact through an energy led economic development strategy?
4. How will the urban and rural communities interact through an energy led economic development strategy?
5. How will energy costs impact the creation of an energy led economic development strategy?

Class Debate

How do the specific Nevada and Michigan proposals align with specific energy led economic development strategies?

Class Writing Assignment

Before any of those big decisions can be made, the GEC Committee needs an energy led economic development strategic plan that outlines:

1. the economic and energy resource profile of Ohio, Michigan, and Pennsylvania;
2. a Strengths, Weaknesses, Opportunities, and Threats (SWOT) analysis for creating an energy led economic development strategy;
3. an outline of the elements in brief of an energy led economic development strategy that can address policy steps needed to promote energy efficiency, recruitment of energy-intensive manufacturing companies and energy company headquarters, infrastructure financing strategies to address the demands of the energy industry, sustainable land use strategies, and strategies needed to create a sustainable energy led economic development approach for the urban and rural centers of these three states; and
4. the format of the energy led economic development strategic plan needs to include sections on the goals, strategies, objectives, and tactics needed to achieve economic success.

The class writing assignment is to draft a five-page, double-spaced energy led economic development strategic plan that addresses the three areas specified above.

Additional Readings

Arcelus, F. J., "An Extension of Shift-Share Analysis," *Growth and Change*, 15 (1984): 3–8.

de Leon, Erwin and Elizabeth T. Boris, *The State of Society: Measuring Economic Success and Human Well-Being* (Washington, DC: The Urban Institute, 2010).

Evans, Keith, *Common Sense Rules of Advocacy for Lawyers: A Practical Guide for Anyone Who Wants to Be a Better Advocate* (Washington, DC: Capitol Net Inc., 2010).

Gorden, Gerald L., *Strategic Planning for Local Government* (Washington, DC: ICMA Press, 2005).

Jannson, Bruce S., *Becoming an Effective Policy Advocate, from Policy Practice to Social Justice*, 6th ed. (Independence, KY: Brooks/Cole Cengage Learning, 2010).

King, Bert T. and L. Irving, "Janis, Comparison of the Effectiveness of Improvised versus Non-Improvised Role-Playing in Producing Opinion Changes," *Human Relations*, 9 (1956): 177–186.

Miles, Mike E., Richard L. Haney, Jr., and Gayle Berens, *Real Estate Development Principles and Process*, 2nd ed. (Washington, DC: Urban Land Institute, 1996).

Porter, Michael E., "Location, Competition and Economic Development: Local Clusters in a Global Economy," *Economic Development Quarterly*, 14, 1 (2000): 15–34.

Scheufele, Dietram A. and David Tewksbury, "Framing, Agenda Setting, and Priming: The Evolution of Three Media Effects Models," *Journal of Communication*, 57, 1 (2007): 9–20.

Using Employment Data to Better Understand Your Local Economy, Tool 4. Shift-Share Analysis Helps Identify Local Growth Engines (Penn State University, 2013), retrieved from http://pubs.cas.psu.edu/freepubs/pdfs/UA373.pdf.

Chapter 5

Energy Conservation and Efficiency

Chapter Goals

1. Understand how energy efficiency is measured.
2. Recognize the role that energy efficiency and conservation play in economic development.
3. Understand how educating consumers impacts energy efficiency and conservation.
4. Be prepared to discuss how government mandates impact energy efficiency and conservation.
5. Be ready to discuss how government incentives impact energy efficiency and conservation.

Energy led economic development does not just capitalize on new sources of energy but also looks for ways to promote energy conservation and efficiency. Regions and states that address energy conservation and efficiency can lower the cost companies spend on energy and also can promote many of the alternative energy providers that are focused on growing. Energy conservation and efficiency can promote energy security and address global climate change.

States and regions across the United States have launched energy efficiency programs to encourage economic development. Massachusetts adopted demand side management programs to mandate utilities to promote energy efficiency while implementing electric industry deregulation.[1] Nevada's energy efficiency program started at $2,000,000 in 2001 but jumped to over $30,000,000 by 2006.[2] Fort Collins, Colorado adopted an Electric Energy Supply Policy in 2003 that set a 10 percent citywide energy consumption reduction goal by 2012 and a 15 percent peak demand reduction goal through the use of a demand-side management program.[3] Finally, Utah's comprehensive energy efficiency policy set the goal of increasing the state's energy efficiency by 20 percent, before 2015, through a collaborative strategy between utilities and customers.[4]

Defining Energy Efficiency

Before outlining the benefits of energy efficiency and strategies for economic success, the term energy efficiency must be defined. Energy is an input into the production of a product or service. Energy efficiency is the energy services provided per unit of energy input.[5] The energy efficiency of an air conditioner is the amount of heat removed from air per kWh of electricity output.[6] Energy efficiency is a part of the cost of production and energy efficiency can be measured as part of the Gross Domestic Product (GDP).[7] Energy conservation is defined as a reduction in the total amount of energy consumed, but energy conservation may not increase energy efficiency, depending upon how energy services change.[8] Energy efficiency does not guarantee economic efficiency. Energy efficiency comes at a cost. To develop a more efficient production process, companies will generally need to make a financial investment in new equipment, services, or processes.[9] The question of whether or not the investment will pay off, and its expected timeframe, becomes a mathematical equation. The decision to invest in energy efficiency must consider future energy prices, changes in their operating costs connected to energy use, intensity of the product use, and lifetime of their existing equipment.[10]

Public Policy Benefits of Energy Efficiency

Energy efficiency provides public policy and economic development benefits. These benefits range from the direct creation of jobs to the reduction of company costs, which leads to great productivity and profitability and provide strong community and social benefits.

Energy efficiency programs are job creators. A $1,000,000 investment in building efficiency improvement creates 20 jobs while the same investment in the economy in general only creates 17 jobs.[11] Nearly 1,000,000 American jobs are linked to the energy efficiency industry.[12] Other than energy efficiency related jobs, energy efficiency investments produce substantial consumer savings and free companies to utilize this new funding on other ventures. Energy efficiency jobs pay above average wages and require less than a college or university degree, offering an economic opportunity much like manufacturing did for many Americans. These energy jobs are also a growing sector of the economy. The Department of Energy estimates these Green Energy jobs will double in number by 2020.[13] This shift in spending supports 17 jobs per $1,000,000 invested in energy efficiency compared to 10 jobs generated through $1,000,000 in spending on energy generation and distribution.[14]

Investments in energy efficiency bring a substantial economic benefit. A six-state study illustrated the fact that a $16,800,000,000 electric energy efficiency investment would increase economic activity by $162,000,000,000, based upon consumer savings, and provide nearly a $100,000,000,000 in contributions to the regional Gross State Product, providing the majority of this funding in reduced costs to consumers just for electric energy efficiency efforts.[15] A natural gas energy efficiency investment of $4,100,000,000 would increase regional economic activity by over $50,000,000,000, boosting the Gross State Product by over $30,000,000,000 while creating over 200,000 jobs.[16] A $6,300,000,000 investment in unregulated fuels energy efficiency efforts would produce over an $80,000,000,000 economic return, adding over $50,000,000,000 to the Gross State Product and creating over 400,000 jobs.[17] In total, the 15-year savings and economic gain from energy efficiency just in a handful of New England states would provide $162,000,000,000 in economic benefit.[18]

Customer participation in energy efficiency programs is on the rise. American ratepayer-funded energy-efficiency programs grew from $2,600,000,000 in 2006 to $7,000,000,000 in 2011.[19] Also, states with energy-resource standards grew from 8 in 2006 to 24 by 2012.[20] 80 percent of consumers are aware of energy efficiency benefits.[21] The benefits of energy efficiency are just starting—upgrading lighting, appliances, electrical devices, and installing better insulation in American homes could reduce the US residential-energy demand by 20 percent.[22] Implementing energy efficiency could create more than $1,000,000,000,000 in economic value by 2020.[23] The stakes are significant: more than $1 trillion in value through 2020.[24]

Energy efficiency investments also have substantial community development benefits. The average American household spends 4 percent of their income on home energy, but lower income American's spend 17 percent of their annual income on energy expenses.[25] Energy efficiency investments by lower income Americans provides critical resources for those in most need of funding to pay for housing and other essential needs. Weatherization assistance programs save $2.53 for every dollar invested.[26] Addressing energy conservation and efficiency provides environmental benefits such as lowering greenhouse gas emissions. It can also benefit utility providers by lowering base load and peak demand, reducing the need for additional power generation and distribution investments. Certain energy efficiency programs can also diversify the energy sources used by utilities. However, energy efficiency and conservation programs drive down energy prices through a reduced demand for these services. Lower energy costs can have a major, positive impact on a regional and state economy.

Energy Efficiency and Conservation and the Location Theory of Economic Development

A large part of the economic development strategies of most regions and states is based philosophically, in the Location Theory of economic development. Regions and states address factors that lead companies to choose a specific location. Early Location Theory encourages communities to address critical public infrastructure as the main element impacting the location of jobs. Thus, regions with deep water ports were and are to this day positioned for global trade as logical locations for ships to import and export goods globally. Location is a prime factor that determines the economic success of a region or state.[27]

A logical outgrowth of the Location Theory is that it forces regions and states to address the cost of doing business in their community. Higher costs for companies and their workers can make a region less attractive to new investors. Thus, regions and states aggressively promote their regions as low cost and look to adopt public policy measures that help make that message a reality. Energy is a substantial cost factor for many companies, and not all the regions of the United States are alike when it comes to energy costs.

Big and small companies alike are impacted by the cost of energy. Energy costs are one of the top three business expenses for over one-third of small businesses.[28] Small companies use energy for a range of business functions, and their primary energy costs center on operating vehicles, heating and/or cooling, and operating equipment. Energy costs for larger, industrial firms is an even greater factor when it comes to making company location decisions. Over 80 percent of company executives responding to a corporate site location survey indicated the availability and cost of energy was either a very important or important factor when considering where to locate their company.[29] The cost of energy is a top ten corporate site location factor.

Energy Efficiency Targets

To determine who best to create energy conservation and efficiency programs, regions and states must understand how Americans use energy. Energy consumption centers primarily on buildings, machinery and equipment, and transportation. As an example, Americans consumed 3,856,000,000 kWh of electricity in 2011, which is more than 13 times greater than its electricity consumption in 1950.[30] As figure 5.1 illustrates, the share of electricity usage by sectors is roughly even between residential and commercial, with industrial direct use lagging behind in third place.

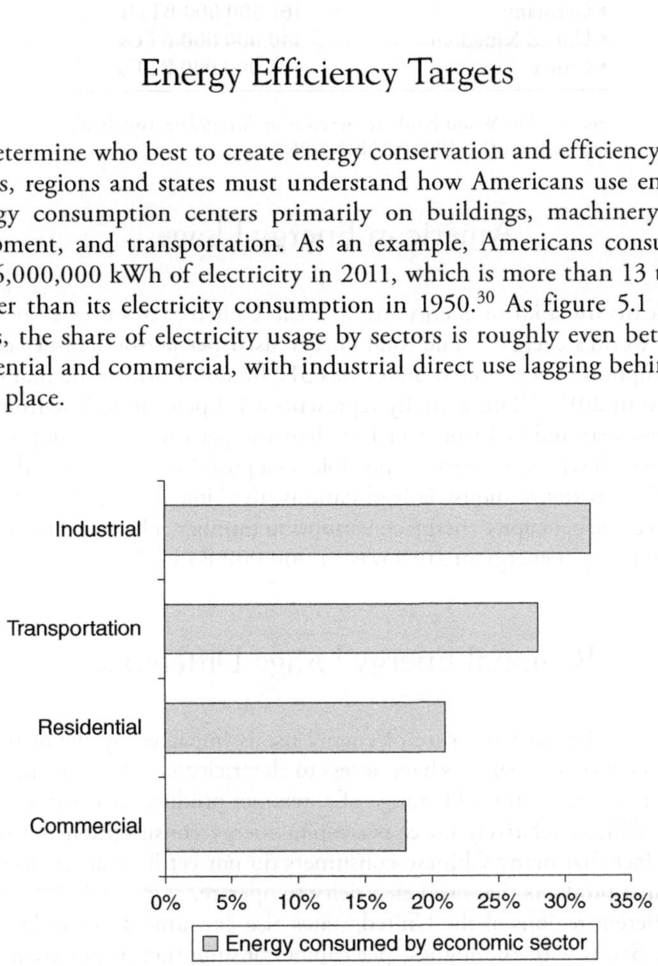

Figure 5.1 Share of total energy consumed by economic sectors.

Source: US Department of Energy, Energy Information Agency, Delivered Energy by Sector, December 2013, retrieved from http://www.eia.gov/forecasts/aeo/er/early_consumption.cfm.

Table 5.1 Industrialized nations' energy consumption per capita basis

• Canada	389,000,000 BTUs
• United States	308,000,000 BTUs
• Japan	161,000,000 BTUs
• Germany	161,000,000 BTUs
• United Kingdom	140,000,000 BTUs
• China	65,000,000 BTUs

Source: The World Bank, retrieved from http://data.worldbank.org/indicator/EG.USE.PCAP.KG.OE.

American Energy Usage

Americans use a lot of energy. In fact, the United States uses 19 percent of the world's energy.[31] The total energy used per person aka per capita consumption in the United States was 312,000,000 British thermal units (BTUs) in 2011.[32] This actually represents a 1.3 percent decline from the previous year and is 13 percent less than the per capita consumption in 1978, which serves, as nearly as possible, as a post-World-War II peak.[33] As table 5.1 illustrates, industrialized nations, the United States has a higher than average per capita energy consumption number. The world per capita consumption of energy in 2010 was 74,000,000 BTUs.[34]

Regional Energy Usage Differences

The per capita basis measure of energy use is impacted by the degree to which a nation's consumers have access to electricity as well as the financial ability to purchase the wide range of consumer products that run on electricity. China's relatively lower per capita energy consumption is a result of the fact that many Chinese consumers do not yet have access to these consumer products that need electricity to operate.

Different regions of the United States also consume different levels of power. States with the highest per capita consumption of power include Wyoming, Alaska, Louisiana, North Dakota, Texas, Iowa, Nebraska, South Dakota, Kentucky, and Indiana.[35] This list of top energy consumers on a per capita basis is dominated either by energy producing states who are home to large industrial complexes that not only produce but also consume a lot of power or rural states whose spread out and smaller energy customer base may increase the overall per capita costs. When electricity consumption is measured in raw total usage, the large industrial states dominate the list of large users. The top states for energy consumption include, in order of rank Texas, California, Florida, Ohio, New York, Pennsylvania, Illinois, North Carolina, Michigan, and Indiana.[36] Weather can also influence not just how much but also which type of energy is used by a state and region. Northern states with cold weather climates use a larger share of natural gas as they need not only to cool their facilities in the summer, but also to heat them in the winter.

Energy Efficiency Investment Decisions

Consumers are faced with a financial decision as to whether an investment in energy-efficient equipment or devices is worth the cost. To make that decision, consumers must understand the different elements of their structures that impact the building's energy efficiency. Addressing energy efficiency often involves impacting the building envelope, which acts as a barrier to air coming in or leaving the structure. A "tighter" envelope improves energy efficiency with fewer gaps through which air flows, keeps conditioned air in. Insulation slows the flow of heat through walls. Addressing the efficiency of the building envelope is a low-cost, high-return energy efficiency tool. Energy efficiency experts also look at the overall performance of a building to address the heating and air conditioning systems, lighting and operation, and maintenance of a structure. Heating and air conditioning systems are commonly referred to as HVAC systems and they include heating, air conditioning, and ventilation systems with air conditioners, heat pumps, furnaces, boilers, rooftop units, chillers, and packaged systems delivered by a system of ductwork and air handlers. 5 percent of the world's energy is consumed by residential air conditioning.[37] Air conditioner efficiency metrics are measured by SEER (seasonal energy efficiency ratio) and EER (energy efficiency ratio) ratings, with higher numbers equating to more efficient air conditioners.[38]

Heating is the largest energy cost in most homes and constitutes up to 50 percent of Americans' annual energy bills.[39] Boilers heat residential and commercial buildings using mostly hot water, but occasionally steam and AFUE (Annual Fuel Utilization Efficiency) are the metric for residential boilers—again, the higher the AFUE, the better the boiler.[40] Thermal efficiency is the metric for the efficiency of commercial boilers.[41] Energy experts begin to suggest the replacement of a furnace or boiler if the units are more than 20 years old. As an example, gas furnaces without an electronic switch generally operate with a gas pilot light and operate at a 65 percent efficiency rating, compared to the over-80 percent efficiency rating of modern gas furnaces.[42]

Energy experts also recommend common maintenance steps that will improve energy efficiency without even replacing the unit. These energy efficiency maintenance steps include cleaning or replacing air filters regularly; cleaning registers; keeping baseboards and radiators clean and unrestricted by furniture, carpets, or drapes; bleeding out trapped air from hot water radiators; tuning up the heating system; sealing the heating system ducts; and checking for wasted fan energy.[43] Monitoring thermostat usage is an energy efficiency tool. 2 percent of a heating bill can be saved for each degree the thermostat is lowered for just 8 hours each day.[44] As an example, turning down the thermostat from 70°F to 65°F saves 10 percent on heating costs, and programmable thermostats help achieve substantial energy efficiency savings.[45]

Water heating is another major energy cost driver. Water heating consumes 12 percent of residential energy and 7 percent of commercial energy.[46] Water heating energy-efficient technologies such as gas water heaters, gas and electric tank-less systems, heat pump water heaters, combination space/water heaters, and solar water heaters all improve water heating costs. Water is heated by electric or natural gas, and the efficiency rating of electric water heaters is 0.9 while gas is 0.6.[47]

Energy Usage by Industry Sector

Americans use plenty of energy for both work and play. Four major sectors of energy consumption are measured in the United States. These include:

- Industrial—facilities and equipment used for manufacturing, agriculture, mining, and construction;
- Transportation—vehicles moving people or goods;
- Residential—homes and apartments; and
- Commercial—offices, malls, stores, schools, hospitals, hotels, warehouses, and restaurants.[48]

As America's post-World War II economic boom happened, energy consumption increased, as along with industrial capacity, transportation networks, and financial means to purchase energy-consuming products. America's growth in manufacturing following World War II and this industry's increased reliance on machines over people created an increased dependence on energy to meet productivity mandates. The development of America's interstate highway system provides the basis for the large transportation sector energy consumption cost.

Finally, Americans rise in per capita income over the last 60 years and the emergence of more consumer products that run primarily on electricity push consumer-based energy consumption up. American energy consumption nearly tripled from 1949 through 2012. Energy consumption increased over the previous years, in all but 18 of the years between 1949 and 2012. Decreases in individual years for energy consumption did happen, but this was generally the result of a slowing economy. Global industrial use of energy constitutes just over half of all global energy use while only constituting one-third of America's energy consumption. Globally, transportation only consumes 20 percent of the world energy but in America, this category consumes 27 percent. America's commercial and residential sectors consume slightly above the global average in energy consumption for these categories.

Industrial Sector

Americans highly productive industrial sector uses less energy than the global average but the nation's expansive transportation network and wealthy consumers promoting individual automobile use drive up American transportation energy consumption in comparison to global counterparts. The industrial sector includes everything from manufacturing of food, paper, chemicals, refining, iron and steel, nonferrous metals, nonmetallic minerals, and other products, and nonmanufacturing generally includes agriculture, mining, and construction. The industrial sector uses energy for processing, assembling, producing steam, cogeneration, heating, air conditioning, and lighting.

More recently, the American manufacturing industry is using less energy. From 2002 to 2010, total energy consumption in the manufacturing sector decreased by 17 percent while manufacturing gross output decreased by only 3 percent.[49] This represents a substantial decline in the amount of energy used per unit in American manufacturing based upon improvements in energy efficiency and changes in the manufacturing output mix.[50] Energy for manufacturing is used either as fuel or as a feedstock (material input to a final product).[51] Five manufacturing subsectors use over 80 percent of the nation's energy fuel: primary metals, petroleum, coal, food, and paper.[52] Only petroleum, coal, and food showed a slight increase in their fuel consumption from 2002 to 2010, but employment in those and all manufacturing sectors fell during the same time period, illustrating the fact that American manufacturing is a smaller but more productive part of the American economy.[53] A review of these heavy industries indicates that 26 separate industrial processes within those industries can have a major impact on energy consumption. If these targeted processes just utilized the "best practices" for energy efficiency, companies in these heavy industries would gain a 26 percent savings from power costs.[54] For nations as a whole, this energy efficiency investment in these heavy power users would constitute 15–20 percent in power savings to industrialized nations and nearly 35 percent in savings to developing nations.[55]

Local governments, while not thought of as a traditional industrial organization are substantial users of energy. Municipal water treatment systems are the largest energy consumers for local government and they account for up 40 percent of their total energy use. It is estimated that energy efficiency measures at the nation's water and wastewater treatment facilities can save up to 30 percent on energy costs.[56]

Residential Sector

America's energy consumption at home changed as well. American homes are bigger but more energy efficient than older homes. American homes are 30 percent larger than their predecessors but consume only 2 percent more energy than older homes.[57] Newer homes are more likely to include modern appliances such as dishwashers, clothes washers, clothes dryers, two or more refrigerators, computers, and multiple televisions that drive higher energy usage. However, new homes are run by far more energy-efficient devices, are insulted better, and have newer, double-pane windows that are more energy efficient. The average American home consumed 90,000,000 BTUs in 2009 and illustrated a continued decline in residential energy consumption while the size of the average American home increased.[58] Homes built after 2000 constitute about 14 percent of all occupied housing units and these new homes consumed 21 percent less energy for space heating on average than did older homes.[59]

Of course, geography played a role as well. Household energy expenditure varies by state, more often than not based upon the weather of the region that triggers a demand for heating and cooling systems. As an example, recently, the annual average New Jersey household energy expenditure was $3,065, which is double the average expenditure for California homes.[60] This Southern movement added to power costs for air conditioning. With 53 percent of newer homes built in warmer, Southern climates, cooling bills are higher in the Southern residential market.[61]

For decades, heating and cooling constituted more than half of all residential energy consumption. In 1993, heating and cooling constituted 58 percent of residential energy consumption, but that number dropped to 48 percent by 2009.[62] Energy-efficient equipment, better insulation, more efficient windows, and population shifts to warmer climates are all credited with this drop in heating and cooling consumption.[63] However, while heating and cooling consumption declined, energy consumption for twenty-first century appliances and electronics continues to rise. 52 percent of residential energy consumption is now covered by non-weather-related appliances, electronics, water heating, and lighting.[64] This number is up by 10 percent since 1993.[65] In total, the average US household consumed 11,320 kWh of electricity in 2009.[66] Americans' newer, larger homes are in fact consuming 19 percent more energy than those built in the 1980s.[67] Heating homes takes up 63 percent of natural gas consumed in US homes, with the remaining natural gas used for water heating, cooking, and other uses.[68]

Different appliances use different amounts of electricity. Appliances that run on a full-time basis, such as heating and cooling units, burn more electricity on an annual basis than those that run on an intermittent basis, such as televisions and computers.

Transportation Sector

America consumes a substantial amount of energy transporting people and goods by road, rail, pipeline, and water. America's transportation network spreads across all 50 states, driven in large part by the interstate federal highway system: local and state roads connecting cities and neighborhoods across the country. It is no secret that America loves cars. The United States ranks third on a per capita basis for nations with the highest use of road motor vehicles per 1,000 residents.[69] Other industrialized nations fall far behind this American ranking. Japan is ranked seventeenth, Germany is twentieth, and the United Kingdom is ranked thirtieth.[70] China is far behind, ranked seventy-first out of just over 180 countries across the globe.[71]

Energy use in the transportation sector is very much impacted by individual travel behavior, land use development patterns, and the extent to which urbanization creates a concentration of customers to make mass transit more financially appealing. World transportation energy usage forecast shows a planned, slight increase in the transportation sector.[72] Petroleum and other liquid fuels dominate the energy sector used for transportation and their forecast growth will occur as well.[73] As with economic growth forecasts, developing countries are expected to see the most growth over the next 30 years. It is no surprise that of this group, China and India will see the largest growth in energy consumption in the transportation sector, with growth rates expected of 4.1 percent and 4.6 percent respectively, while transportation energy per capita usage actually declines in the United States and Europe.[74]

The United States is the largest consumer of transportation energy in the world but its current 27 percent of the world total in 2010 is expected to decline to 19 percent by 2040, based primarily on federal government-mandated fuel efficiency standards, particularly in popular light duty trucks.[75] Of course, the American transportation sector is served almost exclusively by petroleum. 92 percent of all American transportation runs on petroleum and, following the Energy Crisis of the 1970s, petroleum is not used very much for other energy purposes.[76]The American transportation sector is dominated by the traditional family automobile or light truck. These vehicles consume nearly 60 percent of the transportation energy used.[77] Medium to heavy trucks take 22 percent of the transportation energy consumed with air transportation taking 8 percent, water 5 percent, and buses only 1 percent.

Commercial Sector

80 percent of the American economy is in the service industry. It should not be surprising that the commercial sector of the American economy constitutes a large share of the energy consumption. The offices, retail stores, and service centers consume substantial amounts of energy.

Energy Efficiency Strategies

Energy efficiency programs involve efforts related to the promotion of information, use of economic development incentives, and development of product standards. Information is really about empowering consumers to make the most efficient decision related to energy consumption. A wide range of energy efficiency and conservation tools exist that range from investments in the physical capital of facilities to reduction in the loss of energy or toward making the production of power more efficient. Tax credits and other government subsidies are often used to incentivize companies and individuals to make investments in energy efficiency. Regulatory mandates for the use of alternative energy sources is also a major energy conservation and efficiency tool used to promote the alternative energy industry and diversify energy sources.

Information to Promote Energy Efficiency

Promoting the benefits of energy efficiency is the least intrusive method for government to support energy efficiency goals. Empowering the consumer to gain the benefits of energy efficiency does not require an economic incentive or regulatory mandate, but only an explanation as to the financial benefits of the product, service, or structure, which encourages the consumer to act in their best interest. Information as an energy efficiency tool involves the rating of devices and structures and the audit of devices and structures to measure energy efficiency.

The Energy Star rating system is a prime energy-efficiency rating tool that is all about consumer information. The hope is that an informed consumer will make the wise choice in a competitive market, as it relates to energy efficiency. The Energy Star rating system is designed to empower the consumer. Energy Star is a voluntary US Environmental Protection Agency program designed to promote energy efficiency and save consumers' money, established in 1992 as part of the Clean Air Act. Federal law directs the US EPA to "conduct a basic engineering research and technology program to develop, evaluate, and demonstrate non–regulatory strategies and technologies for reducing air pollution."[78] In 2005, Congress further directed the Department of Energy and the Environmental Protection Agency to develop a voluntary program to promote energy–efficient products and buildings through labeling communication about products and buildings that meet the highest energy efficiency standards.[79]

Energy Star product and service ratings are certified through a regulatory process at the US EPA, through a third party certification program.[80] The US EPA certifies products for home and commercial facilities.[81] Product certification is based upon up-front third party testing in EPA-recognized labs, as well as "off the shelf" testing.[82] New homes are also testified for Energy Star certification.[83] New home Energy Star certification is awarded based upon a predefined package of improvements as well as through certification of a customized package of upgrades.[84]

Commercial buildings require the certification from a licensed engineer or architect that it is eligible for an Energy Star rating, based upon acceptable industry standards meeting each of the indoor environment.[85] The same approach is used for industrial facilities but the professional engineer must also certify that the building meets required environmental compliance standards.[86]

The Energy Star program is a strong example of how information alone, not financial incentives or regulatory requirements, can shape consumer behavior and address energy efficiency. As of December 2012,

$239,000,000,000 in savings for residents and businesses have been achieved in utility bills and 1,900,000,000 metric tons of greenhouse gas emissions have been avoided since the initiative started in the 1990s—over the past two decades.[87]

Another prime energy efficiency program where information is key is the use of energy audits. An energy audit identifies where energy is consumed and how much energy is consumed within an existing facility, and this information can then be used to develop energy conservation measures to reduce energy costs and power costs and improve environmental quality.[88] The energy audit involves energy accounting/bill auditing which is the process of collecting, organizing, and analyzing energy data based upon a kWh measure for electricity and similar data measures for natural gas, heating fuel, water, and sewer bills.[89] Energy audits center on an analysis of typical energy bills to determine if a customer is spending more or less on power than similar users do.[90] Measures of windows, water heaters, boilers, HVAC, and other major power users are taken to determine if energy saving investments are worth the up-front cost of energy savings equipment and processes.[91]

Measures of whether an energy audit-suggested remedy makes economic sense are determined in a variety of methods. The Simple Payback method measures how long it takes to recover an initial investment in a cost-saving measure.[92] However, this approach fails to consider the time value of money and the impact of a loss of cash flow. The Discounted Payback model resolves the time value of money issue by discounting future cash flows to their present value and determining the length of time that it takes for the cumulative present value of savings to equal the investment cost.[93] The Return on Investment approach measures the percentage of the investment cost that will be returned annually by savings.[94] The Internal Rate of Return approach calculates the discount rate that makes the present value of the costs equal to the present value of the savings—the Internal Rate of Return needs to be greater than the interest rate at which the money was borrowed to finance the project.[95] Finally, the Net Present Value method discounts costs and or savings at a set rate and then subtracts the resulting present value of the costs stream from the present value of the savings revenue—the Net Present Value must be positive.[96]

The savings from energy audits can be substantial. The Department of Energy's Save Energy Now audit program helped many large US companies, and the average energy cost savings identified with an energy-intensive system is $1,400,000.[97] The average energy savings identified by an audit is 190 billion BTUs (6,900 tce), which constitutes 6.5 percent of the total energy consumed by an audited company.[98] Natural gas savings averaged is 128 billion BTUs (5,300 tce) per energy audit, which constitutes

6 percent of the plant's total natural gas consumption.[99] Small- and medium-sized US companies audited by the US Department of Energy identified $165,000 in average savings per audit and identified, on average, 23 billion BTUs (900 tce) or 8 percent of the total source energy consumed.[100]

The Wyoming Energy Audit Program provides matching grants to qualified small businesses to pay for the energy audits.[101] These state funded energy audits calculate the energy losses due to inefficient equipment and structures, but also provide the cost benefit of investing in energy-efficient improvements and adopting other conservation measures.

Economic Development Incentives to Promote Energy Efficiency

Policymakers wishing to move beyond merely providing information to consumers look to provide financial incentives to encourage energy efficiency. Financial incentives can be awarded by the local, state, or federal government. As table 5.2 shows, they can involve tax credits, tax abatements, grants, rebates on power bills, and low-interest government loans. Taxes impacted include property and sales most often.

Companies are provided a range of tax credits, deductions, and exemptions in many states when they purchase and install targeted renewable energy or energy efficiency equipment.[102] Federal law creates a federal tax credit for residential property owners investing in solar-electric systems,

Table 5.2 Energy efficiency economic incentives

- Corporate Tax Incentives
- Grant Programs
- Green Building Incentives
- Industry Recruitment/Support
- Loan Programs
- PACE Financing Property-Assessed Clean Energy (PACE)
- Performance-Based Incentives
- Personal Tax Incentives
- Property Tax Incentives
- Rebate Programs
- Sales Tax Incentives

Source: Database of state incentives for renewables and efficiency, retrieved from http://www.dsireusa.org/glossary/.

solar water heating systems, fuel cells, small wind-energy systems, and geothermal heat pumps.[103] Many incentives are based upon the level of energy produced by a building and, often, minimum investment levels are required.[104] However, these corporate incentives cap the total incentive available per company as well as have an overall budget cap for the total amount of the incentive available in the state.[105]

Grant programs also promote energy efficiency investments. These focus primarily on funding the purchase of a range of technologies for commercial, industrial, utility, education, and government customers.[106] Grant programs pay down the cost of energy efficiency equipment but some also focus on funding for energy efficiency research and development commercialization.[107] Green buildings incentives are a prime source of energy efficiency funding. Government provides financial incentives for the design and construction of buildings that conserve energy beyond traditional buildings.[108] The most popular economic incentive to promote the development of Green Buildings is the reduction or waiver of a local government building permit fee.[109]

The federal government offers a major incentive for homeowners to finance energy efficiency improvements through an energy-efficient mortgage. The Federal Housing Authority allows lenders to add up to 100 percent in energy efficiency improvements to an existing mortgage loan.[110] These mortgages were limited to $8,000 but the cap is now less than 5 percent of the value of the property, 115 percent of the median area price of a single-family dwelling, or 150 percent of the Freddie Mac conforming loan limit.[111]

The federal government also offers Qualified Energy Conservation Bonds used by state, local, and tribal governments to finance energy projects. $3,200,000,000 in these bonds were issued by the American Recovery and Reinvestment Act of 2009, as allocated by state governments or large local governments.[112] Qualified energy conservation projects include energy efficiency capital expenditures in public buildings that reduce energy consumption by at least 20 percent, green community programs, renewable energy production, research and development, mass commuting facilities, energy demonstration projects, and public energy efficiency education campaigns.[113]

Many local and state governments use economic development incentives targeted at Green companies through the use of tax credits or abatements as a means to recruit companies in the energy efficiency industry.[114] These performance-based incentives are awarded, like most of these incentives, on a competitive basis and only if the company performs the job creation and capital investment promised. The goal behind these incentives is not only to lower the cost of doing business by reducing energy costs, but also to build a new generation of manufacturers, research and development facilities, and corporate headquarters of high-wage, high-growth companies in the renewable energy industry.

Low-interest government loans offer financing to purchase energy efficiency systems or equipment for what is often referred to as demand-side management programs for electrical utilities.[115] State governments also offer low-interest loans for residential, commercial, industrial, transportation, public, and nonprofit organizations around a specific project.[116] Much like gaining a loan from a bank, low-interest government loans offer competitive rates and terms for organizations that may struggle to gain financing from traditional, private sector sources. However, that does not mean the government is looking to lose money. Many of these low-interest government loan programs are revolving in nature and, thus, need repayment of loans to continue to operate the program. Property-Assessed Clean Energy (PACE) financing offers another method for property owners to fund energy-efficiency investments.[117] PACE financing is repaid through a special assessment of the owner's property.[118] Local government also offer production incentives to provide cash payments based on the number of kWh or BTUs generated by a renewable energy system.[119]

Utilities have a range of other methods by which they can meet energy efficiency goals. A Systems Benefits Charge Model requires the utility to create a fund based upon a charge to customers.[120] The utility company, government, or a third party can then administer the fund to provide a financial incentive for customers to make energy-efficient investments.[121] California, Rhode Island, Connecticut, and Massachusetts operate a Systems Benefits Charge Model operated by the utility company while New York, Vermont, and Wisconsin operate a similar program operated by a third party.[122]

Individuals are offered a range of economic incentives to make energy efficiency investments, primarily to subsidize these costly investments. Often taking the form of a tax credit, individual energy efficiency incentives are limited in the amount of the credit that can be taken as well as in the overall budget funding.[123] Credits maybe carried over to future years when an individual's tax payments are higher, or they may even be refundable, which means a tax credit would be given even if the individual does not pay taxes that year. In addition, targeted technologies and forms of renewable energy technology may also be limited by these tax credit programs.

Property tax abatements have traditionally been an economic development incentive used to reward job creation and capital investment. Often referred to as an Enterprise Zone, property tax abatements are among the widest used of all economic development incentives. It is not a surprise then that property tax abatements are also used as an incentive to reward residents or companies for their use of energy-efficient equipment or for alterations to structures in order to promote energy efficiency. Energy efficiency property tax incentives do not tax the added value of a renewable energy system or the increased value of the property based upon the energy-efficient investment in their structure.[124] Thus, the higher cost of a renewable heating system and the subsequent increased value of the property because of the renewable heating system would not be taxed. Local governments implement property tax abatement programs, often permitted through state statute.

Rebates are another popular program with states: utilities and local governments offer rebates offering direct payments to users of energy efficiency systems. Utilities generally administer energy efficiency rebate programs, and the incentive varies by technology.[125] State and local governments provide rebates typically through a sales tax exemption for the purchase of energy-efficiency measures.[126]

Energy Efficiency through Government Regulation and Mandate

Governments promoting energy efficiency and, often, the recruitment of renewable energy companies, will not just stop with the education of their consumers and economic development incentives to encourage the use of energy-efficient measures. As table 5.3 illustrates, these local and state governments push even further and mandate the use of energy efficiency systems, through either consumers or the utilities, with regulatory mandates. The regulatory mandates generally address the efficiency of electricity-powered devices and buildings.

Table 5.3 Energy efficiency mandates

- Appliance/Equipment Efficiency Standards
- Building Energy Codes
- Energy Efficiency Resource Standards (EERS)
- Energy Standards for Public Buildings
- Equipment Certification Requirements

Source: Database for state incentives for renewables and efficiency, retrieved from http://www.dsireusa.org/glossary/.

The establishment of minimum efficiency standards for appliances and equipment is a tool many states use to promote energy efficiency.[127] Appliances and equipment failing to meet these energy efficiency standards cannot be sold at retail centers.[128] These state energy efficiency standards are modeled after similar federal government standards, and the federal standards will preempt the state standard for the device even if the state standard is stricter.[129]

State building codes are the prime regulatory mandate used to promote energy efficiency.[130] All states and larger local governments adopt building codes to ensure all structures are safe for habitation. Many states and local governments' focus on renewable energy as a source of jobs will also adopt energy codes requiring commercial and residential construction to meet certain energy standards.[131] These state and local energy codes are based upon national industry standards adopted by groups such as the International Code Council and the American Society of Heating, Refrigerating and Air-Conditioning Engineers.[132] A handful of states and local governments have gone so far as to require the complete use of Green Building Standards so the structure meets an LEED certified level.[133]

State regulation of utilities for energy efficiency affects its implementation. Energy Efficiency Resource Standards mandate utilities to meet specific targets for energy savings according to a set schedule.[134]

Energy Efficiency Resource Standards establish energy reduction targets for electricity sales, peak electricity demand, and natural gas consumption.[135] Under this regulatory mandate, utilities develop demand-side management programs to provide financial incentives to customers to use energy-efficient equipment.[136] Utilities can also be required to implement energy efficiency strategies, through the adoption of an Integrated Resource Plan in which the utility receives a shareholder incentive, recovery of lost revenue, or other methods of cost recovery for additional expenses related to the implementation of energy efficiency programs.[137] Nevada, Arizona, Minnesota, and the utilities of Bonneville Power Administration, Vermont Gas, and Keyspan all offer energy efficiency through an Integrated Resource Plan.[138]

Government also leads by example when it comes to promoting energy efficiency by adopting strict energy standards for the renovation and construction of their own buildings. Government structures for regions focused on the renewable energy industry will mandatorily be built to meet a certain level of LEED certification.[139] Governments can also mandate their own use of energy-efficient equipment. This can take the form of requiring that all government equipment meet an Energy Star rating or that transportation equipment be modified to operate as efficiently as possible.[140]

Big Questions

1. Is there enough economic value to justify government involvement in energy efficiency and conservation?
2. Do costs of energy efficiency and conservation tactics exceed the economic benefits?
3. Can education alone meet energy efficiency and conservation goals?
4. Are government mandates the answer to meet energy efficiency and conservation goals?
5. Are government incentives really the answer for meeting energy efficiency and conservation goals?

Class Writing Assignment

Remember, Las Vegas is trying to become a global manufacturing center and is looking to reduce energy costs through an energy efficiency program. Outline what should constitute the Nevada Energy Efficiency program and included elements related to consumer education, efficiency incentives, and efficiency mandates in a one-page policy brief. Make sure to start by outlining the benefits of energy efficiency for the state of Nevada and make sure the document advocates for each element of the Nevada energy efficiency plan. The one-page policy brief should be formatted with three or four bold headlines that advocate clear positions followed by three or four bullet points. Manufacture Vegas expects the policy brief to introduce the topic of energy efficiency, define the specific economic and environmental benefits of energy efficiency, and advocate a specific energy efficiency strategy for the state—all in one page.

Class Debate

Nevada State Senator Chip Smith introduced legislation mandating the creation of a state energy efficiency program that includes consumer education, consumer incentive, and utility mandate provisions. The mandate provision requires state electric and natural gas utilities to improve the energy efficiency of the state's business customers by 15 percent through a $100,000,000 rebate program to come directly out of utility profits. The electric and natural gas utilities are opposed to this; the state's manufacturers are supportive, as are the environmentalists. The Nevada Senate Public Utilities Committee will hear testimony, ask questions, debate the bill, and either pass or kill the legislation.

Additional Readings

Frankel, David and Humayun Tai, "Giving US Energy Efficiency a Jolt," McKinsey, December 2013, retrieved from http://www.mckinsey.com/insights/energy_resources_materials/giving_us_energy_efficiency_a_jolt.

Gillingham, Kenneith, Richard G. Newell, and Karen Palmer, "Energy Efficiency Economics and Policy," *Annual Review of Resource Economics*, NBER Working Paper No. w15031, June 2009, retrieved from http://papers.ssrn.com/sol3/papers.cfm?abstract_id=1413595.

"Global Industrial Energy Efficiency Benchmarking: An Energy Policy Tool, United Nations Industrial Development Organization," Working Paper, November 2010, retrieved from http://unido.org/fileadmin/user_media/Services/Energy_and_Climate_Change/Energy_Efficiency/Benchmarking_percent20Energy_percent20Policy_Tool.pdf.

How to Conduct an Energy Audit: A Short Guide for Local Governments and Communities (Trenton, NJ: New Jersey Department of Environmental Protection Division of Science, Research and Technology Bureau of Sustainable Communities and Innovative Technologies, 2006).

Howland, Jamie and Derek Murrow, "Energy Efficiency: Engine of Economic Growth," ENE Environment Northeast, October 2009.

Schantz-Feld, Mali R. "Top Site Selection Factors: Energy Availability and Costs—Reconciling Price and Reliability," Area Development, November 2011, retrieved from http://www.areadevelopment.com/EnergyEnvironment/November2011/site-selection-factors-energy-costs-88872252.shtml.

Thisse, Jacques-Francois, "Location Theory, Regional Science and Economics," *Journal of Regional Science*, 27 (1987): 519–528.

US Department of Energy, Energy Information Agency, American Energy Outlook, McKinsey Analysis, 2008.

US Department of Energy, Energy Information Agency, State Profiles, 2014, retrieved from http://www.eia.gov/state/?sid=US.

US Environmental Protection Agency, Clean Energy Report, Chapter 6, Energy Efficiency Best Practices Report, 2014, retrieved from http://www.epa.gov/cleanenergy/documents/suca/napee_chap6.pdf.

Yusuf, Juita-Elena and Kathaine A. Neill, "State Energy-Based Economic Development Policies and Examples," *Economic Development Quarterly*, 27, 3 (2013): 246–247.

Chapter 6

Energy as Critical Infrastructure

Chapter Goals

1. Understand what constitutes infrastructure.
2. Recognize the connection between infrastructure and economic development.
3. Be prepared to discuss the role energy plays as a piece of a community's infrastructure.
4. Be ready to discuss how the cost of energy impacts economic development.
5. Understand how the utility regulatory and deregulatory models impact economic development.

A Pacific Gas and Electric natural gas pipeline exploded and killed eight people in the San Francisco Bay region.[1] Federal regulators determined that the explosion was triggered by an electrical problem at a control center and inadequate welds in the pipe.[2] The result was 47,000,000 cubic feet of gas escaped into a ball of fire.[3] This explosion is just one example of the challenges faced by America's energy infrastructure. The provision of reliable energy is a critical piece of infrastructure that enables economic development.

California was also facing an energy crisis from a cost standpoint. The California Public Utilities Commission permitted Pacific Gas & Electric to give commercial customers a 12 percent reduction in their electric bills for five years.[4] This electric program will provide even greater incentives in regions with an unemployment rate at 25 percent higher than the state average.[5] These struggling regions gain a 30 percent electric rate reduction.[6] The reduced electric economic development rate is also targeted at high energy use companies with power loads of least 200 kW considering moving out of state or closing due to high energy costs.[7] California's electric rate incentive program illustrates the role the provision of power plays in the economic growth of every state and region.

The cost of energy impacts economic development. Energy-intensive industry sectors such as chemical, aluminum, cement, iron, steel, paper and pulp, glass, and refining account for millions of high-wage manufacturing jobs for working class Americans. In a chemical factory, energy can constitute 80 percent of the company's costs and lower energy costs in 2012 provided a $130,000,000 savings for American manufacturing companies.[8] These energy-intensive industries account for 70 percent of the global energy usage. In the United States or globally, the price of energy affects the economic development of most regions. America's energy infrastructure is generally reliable.

Role of Infrastructure in Economic Development

The Roman Empire would not have governed the world for so long without the incredible infrastructure developed to serve their people, businesses, and military across Europe. Rome's system of aqueducts provided the extra water that the Tiber River could not for the citizens of Rome, but also ensured the water supply was clean to prevent the illnesses that followed many other less advanced civilizations.

Today is no different. Economic development does not happen without the essential physical infrastructure needed to retain and attract the organizations that keep an economy running. Infrastructure is the substructure or underlying foundation or network used for providing goods and services.[9] The manufacturing industry in particular is dependent upon community infrastructure to grow their industry.[10] Public infrastructure investments reduce the cost of doing business for local businesses as they would have to invest in the roads, water, sewer, power, and other infrastructure needed to operate in a global economy. Infrastructure investments also eliminate the challenges for isolated rural communities by linking them with larger, more productive urban markets. The economic success of a region is determined not just by how close the region is to major economic centers but how quickly customers and workers can be linked to those economic centers. This investment in infrastructure benefits urban and suburban centers and eliminates the economic isolation of rural communities.[11]

Economic development cannot happen without public infrastructure investments. However, as figure 6.1 illustrates, infrastructure is not just roads, water, and sewer lines.

Figure 6.1 Infrastructure example.

A broader perspective of what constitutes infrastructure is essential if regions and states are to succeed in retaining and attracting jobs. The provision of infrastructure traditionally is the role of government as an essential public service as well as an economic development tactic. Productivity of employees and companies is increased with infrastructure investments.[12] These infrastructure investments also reduce the cost of doing business for companies and make the region and state more attractive.

Energy as Infrastructure

Energy is an infrastructure issue influencing economic development. The provision and cost of energy impacts the location of companies and the quality of life for the region and state. The availability of low-cost, reliable power is another essential piece of physical infrastructure for any development. Computers, lights, air conditioning, heat, and nearly every piece of equipment and machinery in a facility need power. The Industrial Revolution was possible because electricity enabled the use of a new generation of machinery in nineteenth century factories. The Information Age is no different, as information technology is dependent on low-cost, reliable power. America's energy infrastructure is under scrutiny. Nationally, energy demand is expected to grow by 8–9 percent going into 2020.[13]

America's electrical grid and pipeline distribution systems still operate with large portions constructed in the nineteenth century. Complicating matters further are regulatory permitting issues, weather events, and limited maintenance in a growing and deregulated market, thus increasing the number of power failures and interruptions.[14] Of even greater concern is that after 2020, population growth will drive the electricity capacity beyond current transmission capacity.[15] 150,000 miles of crude oil pipelines and over 1,500,000 miles of natural gas pipelines are underground crossing multiple states.[16] Multiple oil and gas pipeline failures have led to deaths, injuries, and property and environmental damage.[17]

Energy is consistently a top ten factor impacting corporate site-location decisions. Corporate site-location consultants rank energy availability and costs as the ninth most important factor affecting site location decisions.[18] 82.1 percent of survey respondents ranked this category as "very important" or "important."[19] Typically, the ranking of energy and its costs rises or falls based upon the price of energy for a given year, but it is consistently an issue impacting economic development, and an essential piece of community infrastructure.

- The cost and reliability of energy for a firm's transportation and logistics services matter as well. America consumes 27 percent of its energy on transportation and only 20 percent on industrial services.[20] Petroleum fuels 93 percent of America's transportation costs. Addressing the costs and reliability issues for a company's facilities can be negated by skyrocketing transportation costs.[21] A survey of small business owners illustrates the challenges associated with increasing gasoline prices:72 percent claim high gasoline prices directly impact their business;
- 41 percent said high gasoline prices affect their plans to hire new employees;
- 22 percent have cut back employee hours;
- 40 percent have raised their prices; and
- 43 percent believe energy prices impact the ability of their company to survive.[22]

Cost of Energy

Not all regions are alike when it comes to the cost of energy. Regions with higher energy costs are at a disadvantage when it comes to recruiting large energy users. The cost of production simply becomes too high.

As figure 6.2 shows, the lowest cost energy states are often big producers of energy. The average cost of energy is about $18 per BTU with the energy producing states and states in the Midwest, West, and South. The island of Hawaii and the northeastern states dominate the bad side of the energy cost

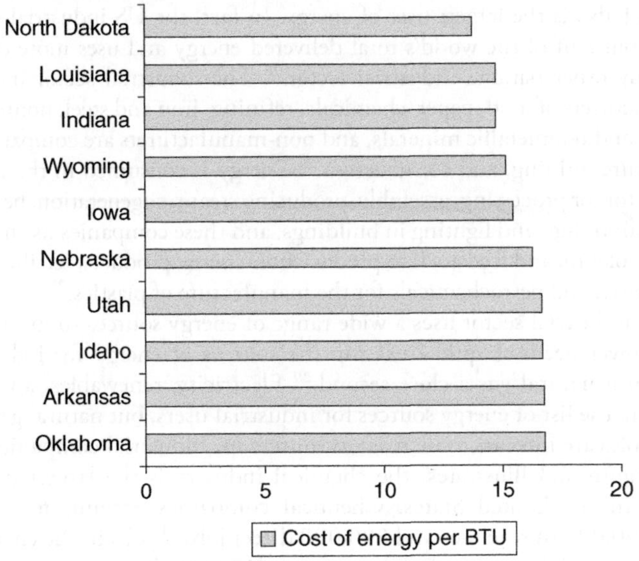

Figure 6.2 Top ten lowest energy costs states.

Source: US Department of Energy, Energy Information Administration, Table E15 Energy Prices and Expenditures Ranked by State, 2012, retrieved from http://www.eia.gov/state/seds/data.cfm?incfile=/state/seds/sep_sum/html/rank_pr.html&sid=US.

list. Many industries are reliant on electricity as their prime power source to operate their facilities. States such as Idaho, Wyoming, Washington, Utah, and Kentucky offer the lowest electricity rates for consumers, and New Hampshire, New York, Alaska, Connecticut, and Hawaii top the list with the highest national electric rates.[23] The cost of transportation energy is worthy of a review as well. The states with the lowest gasoline prices include South Carolina; Arkansas; Oklahoma; Wyoming; Alabama; and Tennessee.[24] Much like facility energy costs, high gasoline prices are dominated by states on the East and West coasts and the island of Hawaii.[25]

High Energy Users Can Be High-Wage Job Producers

Many high-energy users also provide the high-wage jobs that regions and states demand. The industrial sector of the economy provides high wage jobs and also is the largest user of energy. In fact, the US industrial sector uses about half of the world's total delivered energy and uses more energy than any other nation's industrial sector.[26] The industrial sector includes manufacturers of food, paper, chemicals, refining, iron and steel, nonferrous metals, and nonmetallic minerals, and non-manufacturers are companies in agriculture, mining, and construction.[27] Energy is consumed in the industrial sector for processing, assembly, producing steam, cogeneration, heating, air conditioning, and lighting in buildings, and these companies use natural gas, petroleum, and feedstock to produce non-energy products, fertilizers for agriculture, and petrochemicals for the manufacture of plastics.[28]

The industrial sector uses a wide range of energy sources to meet their large power needs. Liquid fuels top the sources of energy for industrial users with natural gas a close second.[29] Electricity, renewables, and coal complete the list of energy sources for industrial users, but natural gas and renewables are forecast to increase as sources for industrial companies.[30]

As figure 6.3 illustrates, the chemical industry is the largest user of energy in the United States. Chemical companies account for nearly $700,000,000 in revenues and 800,000 direct jobs.[31] Jobs in the chemical industry are high wage and often pay over $100,000.[32]

Big data is another big job producer that is addicted to energy. Nearly all industries have massive amounts of data affecting their business. "Big

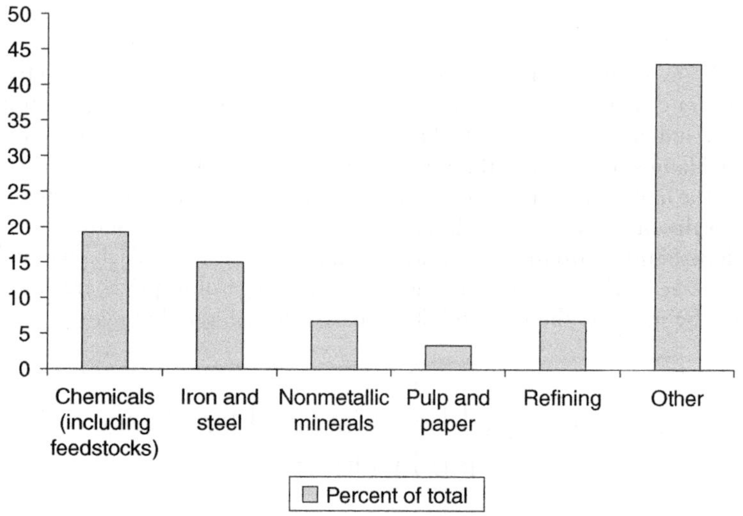

Figure 6.3 Shares of total energy consumption by industry sector, 2010.
Source: US Energy Information Administration, International Energy Outlook 2013.

data" is defined as datasets whose size is beyond the ability of the typical database software tools to capture, store, manage, and analyze.[33] To illustrate the economic impact of Big Data, if the US healthcare industry could use Big Data effectively, a $300,000,000,000 savings could be produced.[34] Big Data's applications are well beyond healthcare as it only costs $600 for a disc that can store all the world's music, 5,000,000,000 mobile phones are in use, 30,000,000,000 pieces of content are shared on Facebook, and 15 out of 17 US industry sectors have more data stored per company than the Library of Congress.[35]

Big Data is a hot topic among economic developers and with good reason. Big Data is expected to create 1,900,000 American jobs and 4,000,000 jobs globally by 2015.[36] Big Data jobs are the high-wage jobs paying in excess of $100,000 for which every community is searching.[37]

Big Data jobs are not easy to get and many obstacles from broadband to workforce must be addressed, but the availability of low-cost affordable power is one of these obstacles. Big Data analytics centers are substantial users of power, much like data centers and other tech intensive office facilities. The cost and sources of power impact the ability of regions to gain the Big Data centers everyone is recruiting.

Data Centers are another major focus for economic developers that spend substantial funds on power. The creation of "cloud" computing is changing the way the Internet operates and influencing the information technology infrastructure needed by world's leading high-tech companies. The "cloud" is creating billion dollar Data Centers bringing tax dollars and jobs to regions across the United States.

- Data Centers are broken down into four types:Tier One—single path for power and cooling distribution with no redundant components and less than 28.8 hours downtime per year;
- Tier Two—single path for power and cooling distribution with redundant components and less than 22 hours downtime per year;
- Tier Three—multiple power and cooling distribution paths, but only one path active and redundant components, concurrently maintainable and less than 1.6 hours downtime per year; and
- Tier Four—multiple active power and cooling distribution paths, redundant components, and little fault tolerance with less than .4 hours of downtime a year.[38]

The growth of data centers is expected to be substantial. New construction for substantial data centers in 2010 reached $6,000,000 and is expected to double by 2015, driven by spending on cloud technology booming from $16,200,000,000 in 2008 to over $42,000,000.[39]

- A number of Data Center announcements have recently been made, including:Apple built an iDataCenter in rural Maiden, North Carolina with a 500,000 square foot main building and a 30,000 square foot "tactical" data center next door powered by two solar arrays and a "fuel cell farm" featuring Bloom Energy Servers fueled by gas from a nearby landfill;
- Google built a $1,500,000,000 Data Center in Council Bluffs, Iowa and a $1,200,000,000 data center in Lenoir, North Carolina;
- Microsoft built a billion dollar data center in small and rural Boydton, Virginia;
- The federal government built a billion dollar data center for the National Security Agency in Bluffdale, Utah; and
- Microsoft, Yahoo, Sabey, Dell, Intuit, and Vantage collaborated and built a billion-dollar data center in small town Quincy, Washington.[40]

The economic impact of data centers can be substantial. The Quincy, Washington facility boosted property tax values from $260,000,000 to $764,000,000 and boosted property tax collections by more than $1,400,000 and school taxes by $1,600,000 in this small town over just three years.[41] A similar project in Oregon created 80 jobs.[42] While some communities such as the Google project in North Carolina have mitigated the positive economic effect through aggressive tax incentives, most data centers are a net positive for small rural towns.

Data centers are massive users of power and the availability of low-cost, reliable energy is a top priority when considering where to locate a data center. Data centers can consume up to 100 times more energy than a standard office building of the same size and less than 15 percent of the original source energy is used for information technology equipment in the data center.[43] Data center energy consumption doubled from 2000 to 2006.[44] In 2010, data centers used 1.5 percent off US energy, costing $4,500,000,000 annually.[45]

Reducing Energy Costs

Cheap energy is provided in a range of ways. Natural resources utilized for energy production often provide a low-cost energy option challenging to meet for regions without those resources. Many states utilize either a market-based or regulatory approach to address energy costs. Whether nature or regulators give a company cheap power options, reducing energy costs is a focus for policymakers.

Natural Resources

Most regions of the United States have access to an affordable source of energy based upon the natural resources available in the region. Hydro, geothermal, shale, oil, natural gas, wind, solar, biomass, and other sources of natural resources are available and, with strategic infrastructure investments, can become a substantial source of affordable power.

The Tennessee Valley Authority (TVA) is a twentieth century success story when it comes to harnessing the power of nature to develop low-cost electricity. Born out of a 1933 Act of Congress, the TVA is the nation's largest public power provider and serves most of Tennessee, parts of Alabama, Georgia, Kentucky, Mississippi, North Carolina, and Virginia.[46] This constitutes services to more than 9,000,000 people over 80,000 square miles.[47] The TVA serves 155 municipal and cooperative power distributors and directly serves 57 large industrial or government installations.[48] TVA's operating revenue is approaching $12,000,000,000 and they operate without tax dollars.[49] The TVA generates power from fossil fuels, nuclear and major hydropower plants, and renewable sources.[50] They manage the nation's seventh largest river site to mitigate flood damage, produce power, maintain river navigation, protect the water quality, and provide recreational activities.[51]

The TVA operates a regional economic development program tied to their low-cost, public power offering. The TVA's economic development programs stimulated $5,900,000,000 in business investments recently, recruiting or expanding to more than 150 companies and added 10,750 jobs in East Tennessee, Virginia, Georgia, and North Carolina.[52] 300,000 jobs were retained and attracted by the TVA from 2005 to 2012.[53]

TVA's economic success is driven by the low-cost power they offer to industrial and commercial customers. The TVA offers industrial customers a power rate 16 percent lower than the national average, commercial rates 9 percent lower than the national average, and residential rates that are 20 percent lower than the national average.[54]

A twenty-first century energy success story similar to the TVA is the impact shale developments have had here on the price of energy in the United States. Shale-based oil and natural gas are not only influencing the wealth of regions, but they are also driving down the cost of American energy. It is an issue of supply and demand. The domestic oil and natural gas flooding the market from shale-based drilling sites drastically reduced American gas import needs. Gas imports fell by 32 percent and LNG imports collapsed by 78 percent from 2007 to 2012.[55]

Shale oil resources went from 4,000,000,000 barrels annually in 2007 to 33,000,000,000 in 2010.[56] Production of shale-based oil is dramatically impacting local prices and could lead to oil prices significantly lower than projected current forecasts.[57] The shale boom kept US oil prices lower than global oil costs.

Opponents of shale development point to estimates that shale well production will slow substantially by 2030 and America will be back in its pre-shale world—dependent on foreign oil. Opponents of shale development also point out that regions developing this large pool of shale oil and gas are not guaranteed long-term economic success. They point to the once booming and now empty coal towns in Ohio, West Virginia, and Pennsylvania, as well as the ghost towns dotting the western United States from the nineteenth century gold rush.

Power Regulatory versus Market Approach

The cost and reliability of electric and natural gas service impacts a region and state's economic development. America's industrial, commercial, and residential developments operate primarily on electricity and/or natural gas provided by utility companies authorized to provide service on a state-by-state basis.

For the most part, state policymakers have the largest impact on the price of local electric and natural gas services. State policymakers influence the cost of power for business and residential customers when deciding whether regulators or the free market should set provide retail electric and natural gas service. Figure 6.4 illustrates that the states are evenly divided between regulated and deregulated energy markets.[58]

Regulated Utility Model and Economic Development

Traditionally, utility companies were charged by government with the responsibility to generate, transmit, and distribute power. The government told the utility the rates they could charge and the utility was given an assurance of a monopoly market for customers and typically, a guaranteed rate of return to cover the cost of investments and employees. The consumption of power was measured by meters located at customers' homes or facilities, and charges were based upon a set rate established by state regulators.

Economic development is alive and well in states with regulated utilities. Indiana is a state that is aggressive not just in economic development,

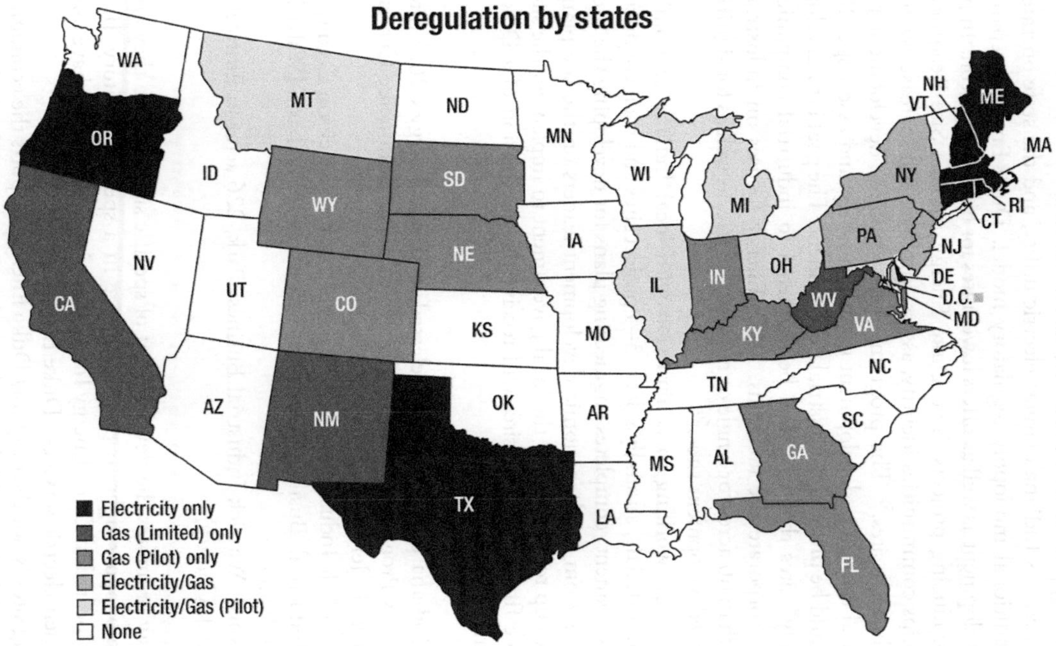

Figure 6.4 Utility regulated versus deregulated 50 state map.

Source: Quantum Gas, retrieved from http://www.quantumgas.com/list_of_energy_deregulated_states_in_united_states.html.

but the state's regulated electric market produces rates below the national average. A prime provider of electric power in Indiana is Duke Energy and they offer an aggressive economic development program. It includes site development support as well as electric rate incentives.

Duke Energy is Indiana's largest electric utility, and the state operates under a traditional monopoly regulatory model. Duke Energy prepares sites for development as well as offers lower rates for high-energy consumption, job-producing projects. The Duke Energy Indiana Site Readiness Program helps communities identify, assess, improve, and increase awareness of industrial sites.[59] The program spurs economic development by assisting local economic development groups to screen and assess specific sites that could be prime for industrial development. The goal is to develop "shovel ready" sites that can then be marketed to industrial companies looking for corporate site locations. Duke Energy focuses on industrial sites larger than 60 acres for single or multitenant use, as well as for significant urban redevelopment sites.[60]

Duke Energy, working in conjunction with local Indiana economic development officials, annually selects 3–5 project sites and then conducts an initial assessment, completes engineering plans for developing the site, gains buy-ins from local economic development leaders on the site plan, and provides up to $10,000 to the local government to implement the site plan.[61] Once finalized, these sites will receive the designation of a Duke Energy "Build Ready" Site.

- Four Indiana properties were selected for Duke Energy's 2013 Site Readiness Program, including:The River Ridge Commerce Center, 100 acres in Jeffersonville (Clark County);
- The East Side Industrial Park, 94 acres in Seymour (Jackson County);
- The Industrial Business Complex, 135 acres in Wabash (Wabash County); and
- The North Warrick Industrial Business Park, 226 acres in Elberfeld (Warrick County).

Duke Energy promotes the development of specific sites because they are granted a monopoly right to provide service in a specific territory. If a company is located in Duke Energy Indiana service territory they have one choice for their electric service—Duke Energy.

Preparing sites is not the only way Duke Energy supports the economic development of Indiana. Duke, with regulatory approval from the state of Indiana, offers electric rate incentives to eligible projects. The Duke Energy Indiana electric rate incentive allows a four year declining reduction (20 percent, 15 percent, 15 percent, 5 percent) on the Duke Energy electric bill for new energy usage associated with an economic development project.[62] The Rider applies to customers with at least 250 hours of use each month. The electric rate incentive must be a stated factor in the company's decision to expand or locate in Indiana, and the customer must also receive a significant state and/or local economic development incentive.[63] The customer needs to have a minimum level of electricity usage with a per project cap of 25,000 kW and a 60,000 kW per year cap, and the new energy usage load must be a minimum of 1,000 kW demand at one delivery point.[64] Job creation is a requirement as well. The customer must create 10 new full time equivalent jobs and invest $1,000,000 per 1,000 kW of eligible new or additional energy usage load or create 25 new full time equivalent jobs per 1,000 kW of eligible new or additional load and invest $8,000,000 per 1,000 kW of eligible new or additional load.[65] A 10-year service agreement is the final piece of the Duke Energy electric rate incentive puzzle.[66]

Finally, Duke Energy focuses on Brownfield redevelopment in Indiana. The Duke Energy Brownfield rider is available until December 31, 2016, to customers in a qualified "brownfield" redevelopment area adjacent to a Duke Energy electric transmission or distribution line with adequate electric supply.[67] Customers developing in an Indiana Brownfield in Duke Energy's territory have an opportunity to gain substantial rate reductions through a graduated scale, over a period of five years, as follows:

- For the first year, the maximum load charge (demand charge) is reduced by 50 percent;
- For the second year, the maximum load charge is reduced by 40 percent;
- For the third year, the maximum load charge is reduced by 30 percent;
- For the fourth year, the maximum load charge is reduced by 20 percent; and
- For the fifth year, the maximum load charge is reduced by 10 percent.[68]

The Duke Energy Brownfield incentive program is an example of a utility focusing special rates on sites that are more difficult to develop.

Utility Deregulation and Economic Development

Deregulation of the power industry is the trend among state governments. Under a deregulated power market, utility companies are forced to divide up their companies into separate entities operating power generation, transmission, and distribution. Generation companies can be traditional utilities or merchant operations.[69] These companies have multiple wholesale customers for the marketing of the energy they develop. Transmission companies are regulated by the Federal Energy Regulatory Commission (FERC) and are required to allow anyone to access their transmission systems.[70] This regulation of electric transmission companies allows equal access to power and permits traditional utilities to own transmission companies.[71] Regulations require utilities to take a passive role in the operation of transmission companies.[72] Distribution companies supervise the wires and the "last mile" of the electric market. These companies are typically regulated by state public utility commissions and are the company's customers call when there are service issues.[73] Finally, deregulated energy markets have retail companies competing for customers. Many of these companies create a competitive advantage through the development of aggregation programs, and they entice new customers with reduced rates.

Ohio, Pennsylvania, New Jersey, Illinois, Michigan, New York, Connecticut, Rhode Island, Maine, New Hampshire, California, Arizona, Nevada, Oregon, Maryland, Delaware, and Texas are considered "deregulated" for electric power purposes while Virginia and Montana recently enacted laws to "reregulate" after policymakers grew concerned about the price impact of deregulation.[74] Under utility deregulation, power service marketers offer business and residential customers a new choice for retail electric or natural gas service. Incumbent utility companies may develop new retail companies and retain generation and distribution companies.

Texas is a leader in the deregulation of utility services, and whether this program is a success or failure depends upon whom you ask. Texas is a national economic development leader, creating more jobs than the other states combined and whose income grew faster than any state but North Dakota. However, Texas, while being home to a large base of energy companies, only offers just below average energy costs. Texas finishes twenty-seventh out of the 50 states for electricity costs and fifth in the nation for electricity consumption.[75] Of course, Texas's high electricity costs are in line with Florida and other warm weather states that live off of air conditioning for several months of the year.

It took a decade, but Texas' electricity rates under a deregulated framework are now below national averages.[76] Electricity deregulation in Texas began in 2002.[77] Texas power customers not served by municipal or cooperative utilities can choose their electricity service from multiple retail electric companies. These companies could be subsidiaries of existing utilities or out of state utilities interested in the Texas market. The incumbent utility still owns and maintains the transmission of power under traditional monopoly regulation. In Texas, power generating companies cannot sell or transmit their power to end users and retail electric providers buy power from the power generating companies and sell power to customers, but might not generate or transmit power. Traditional utilities continue to own the generation and transmission end of the power business, and these services remain regulated.

85 percent of Texas' commercial and industrial customers switched power providers and 40 percent of residential customers in deregulated areas of the state switched from the former incumbent provider to a competitive electric company.[78] Consumer groups are not sold on the benefits of deregulation for Texas electric customers. They claim Texans living in deregulated electricity areas paid about $22,000,000 more in the last decade than they would have under a regulated system and Texas residential consumers paid as much as 45 percent more for deregulated electricity than their counterparts in regulated areas.[79] Similar negative impacts of electric deregulation have been discussed in the export of cheap energy from hydropower from the Columbia River in Washington State.[80]

Ohio offers another interesting case study for how energy regulation influences economic development. Starting in 1999, Ohio deregulated its electric market but froze rates during a "Market Development Period" set to expire at the end of 2008.[81] As the date for the rate freeze expiration began to close in, state policymakers chose to partially eliminate the established deregulation framework by creating a new pricing context, and promoted alternative energy by creating a goal for electric utilities to obtain 25 percent of their energy from alternative sources by 2025.[82]

Ohio operates a hybrid electric regulatory framework that permits investorowned electric utilities to go to the marketplace or opt for a non-market based regulatory framework where regulators set rates. Consumers in Ohio responded to the opportunity to switch electric providers. Incumbent companies, which are permitted to form their own competitive retail service providers, have lost a majority of their customers, as table 6.1 illustrates.

Ohio also offers customers a chance to participate in a traditional electric rate rider program known as a "reasonable arrangement."[83] Electrical utilities and business customers may gain a reduced electric rate for non-retail projects that create 25 full-time jobs over three years with an annual average wage that is 150 percent of the federal minimum wage.[84] Companies must also be awarded local, state, or federal tax incentives.[85] An economic arrangement may also be granted to retain utility customers likely to cease or reduce operations or relocate them out of state but the customer have an average billing load of at least 250 kW, and the cost of electricity be identified as a major factor in the decision to cease, reduce, or relocate operations.[86]

Ohio also offers energy efficiency arrangements to provide an incentive for developing energy-efficient production facilities that create ten

Table 6.1 Summary of switch rates among Ohio incumbent electric utilities

Company	Electric Choice Sales Switch Rates (in percent)
Cleveland Electric Illuminating Company	85.83
Duke Energy Ohio	76.6
AEP Ohio	60.61
Dayton Power & Light	69.9
Ohio Edison	79.73
Toledo Edison	87.17

Source: Public Utilities Commission of Ohio, retrieved from http://www.puco.ohio.gov/puco/index.cfm/industry-information/statistical-reports/electric-customer-choice-switch-rates/#sthash.LwIEBsjM.dpbs.

full-time jobs over three years and meet the same criteria as economic expansion arrangements.[87] Ohio's Development Services Agency offers direct funding to manufacturing companies to cover the cost of energy efficiency programs. Finally, Ohio permits utilities and companies to submit applications for unique arrangements that do not require job creation, threats of closure/relocation, or energy efficiency but require the filing party to bear the burden of proof.[88] Ohio guidelines define how the level of incentives may be judged for all the reasonable arrangements including demand discounts, percentages of total bills or portions of bills, direct contributions, reflections of cost savings to electric utility, shared savings, and a combination of all the above.[89]

The results of Ohio's regulatory approach to utility economic development have been mixed. Several economic development arrangements have been approved by the PUCO thus far. The benefits of the incentives have ranged from $ 30,000,000 to $ 60,000,000; however, Ormet, a major manufacturer in southeastern Ohio, closed its facility claiming the state failed to address their high electricity rates.[90] Ormet had gained several state-based negotiated electricity price reductions and faced substantial labor unrest as well.

Energy Diversification and Infrastructure

America strives for energy independence. A key to energy independence is gaining a large share of the nation's energy from domestic sources. The explosion of new oil and natural gas sources from previously dormant shale deposits creates a unique opportunity for America to diversify its energy usage where it is the most dependent on one source: gasoline, which fuels over 90 percent of America's cars. Shale developments create a market for low cost natural gas to power America's transportation fleet.

America would benefit from diversification of its energy supply. Green advocates land on energy diversification arguments as a way to promote renewable energy sources such as wind, solar, biomass, and other sources. Traditional energy companies are busy diversifying their energy sources, looking for additional domestic sources of energy as the stability of the Middle East and other global markets comes into question. While the renewable and traditional energy leaders rarely agree on other issues, they do agree the national economy would benefit from diversifying its energy sources.

The transportation sector of the American economy is ripe for energy diversification. The transportation sector accounts for nearly 30 percent of all American energy consumption.[91] America's need for speed consumes nearly more than their energy consumption for industrial, residential, and commercial purposes.[92] Of greater concern is the fact that petroleum is the source for 93 percent of America's transportation energy needs, with natural gas and renewable sources covering just 3 percent and 4 percent respectively.[93] The reliance on petroleum is actually down from 96 percent in 2008, but is not yet diversified. The American economy is not only auto-centric, but, 40 years after the Arab Oil embargo, is still highly dependent on oil to operate the economy.

The development of new domestic sources of oil and natural gas from once dormant shale deposits position regions and states with a unique opportunity to diversify their transportation energy consumption. In particular, the growth in domestic natural gas production from shale regions creates a new, cheap source of local natural gas that can revive the Compressed Natural Gas (CNG) vehicle marketplace. CNG vehicle technology is proven, and operates in the United States and throughout the world. However, with just over 1,000 CNG fueling stations compared to over 118,000 gasoline fueling stations, CNG needs infrastructure development to serve additional CNG vehicles.[94]

CNG cannot become a true alternative to gasoline for America's fuel supply unless a larger fueling station infrastructure is constructed. Unlike diesel fuel and gasoline, CNG is a gas and not a liquid and has a separate set of safety and construction issues. There are four separate types of CNG station designs: cascade fast-fill, buffered fast-fill, time-fill fueling, and combination-fill. A cascade fast-fill system is similar to a traditional gasoline system in that it is designed for general retail use, but it must have more than one core compressor to provide a continuous flow.[95] The fast-fill systems consist of a dryer, compressor, priority valve panel, storage, sequential valves, temperature compensation system, and dispenser.[96]

A buffer fast-fill system is designed for transit buses and taxis as they fuel directly from the compressor into the vehicle and use a small amount of storage.[97] These systems are for captive fleets and are able to dispense a large quantity of gas in a short period of time.[98] A time-filling system is a low-cost alternative for fleets that wake up and spend the night in the same place, and is popular with school buses and refuse trucks.[99] Costs are kept down on these systems as there are no priority, storage, and sequential fueling components.[100] The cost for CNG fueling stations ranges from $5,000 for a single vehicle to $1,000,000 for large facilities.[101]

The majority of CNG fueling stations in place now are for private onsite fueling; thus, there is a massive market for growth. Most growth models first center on gaining from the fleet truck market that currently is served by over 7,000,000 diesel trucks.[102]

The availability of new domestic sources of natural gas has the ability to reduce the costs of natural gas, but other issues need to be addressed to promote the development of a CNG fueling station infrastructure. Government regulation and purchasing mandates are the most likely route to developing additional market demand for CNG vehicles targeted through government and fleets of commercial trucks. These vehicles travel a known set of miles, over an identifiable route, and they start and end the day in the same place. Recommendations for expanding the CNG fueling station infrastructure include development of regulatory tariff charges for transportation of CNG, mandating the use of CNG for government sponsored fleets such as transit and school buses, and provision of tax credits for CNG fueling stations.[103]

CNG is just one version of the development of an alternative fuel system to diversify America's transportation energy marketplace. No matter the energy source or the technology involved, no alternative fuel system will be successful without the creation of the infrastructure network needed to fuel these vehicles. Drivers need to know they have readily available options to fuel their vehicles when the gas gauge gets close to empty.

Big Questions

1. How critical is energy as a piece of America's infrastructure?
2. How are different industries impacted by energy from an infrastructure standpoint?
3. Can shale do more harm than good?

4. Is electric and natural gas deregulation a short or long-term experiment?
5. Does a regulated utility market produce true market prices for power and help, or does it hurt economic development?

Class Writing Assignment

Remember, Prime 101 is a substantial Michigan economic development project. Their chemical manufacturing center will employ 200 people paying $50,000 a year for this $75,000,000 facility. This company follows the shale boom around the nation and is considering an investment following the Utica and Marcellus shale boom in Eastern Ohio. Prime 101's connection to shale is the interest in using a by-product of the shale oil refining process, but also the need for cheap energy. The company lacks a connection to either state, but they have asked you to serve as their site selection consultant for the project. First, they asked you to develop a one-page summary of the demographic and economic data for both states based upon census data and an outline of the state's economy and economic development incentives. Next, they asked for you to draft the project introduction letter to economic development leaders in the state of Ohio and Michigan to kick off the project. The letter should outline the importance to economic development that the project bears, how it impacts and relates to the state in question, what are the site needs, and what economic development incentive priorities matter. The letter should not exceed two single-spaced pages.

Class Debate

Remember, the Mayor of Las Vegas's proposal to boost the struggling solar power industry and reduce the region's cost of energy by creating a municipal electric company and developing solar farms for energy? Solar City Power is a controversial proposal, as it puts government in a business that has been private and investor-owned for decades, and it is pending before the Las Vegas City Council. The Mayor, solar industry, and area manufacturers support the proposal, as they like competition in the utility business. The utility industry and local chamber of commerce and community groups are opposed to the creation of a government run utility. The City Council will introduce the resolution authorizing the creation of the Solar City Power municipal electricity company, but where the other council members are is unknown. Proponents and opponents will testify and a vote on the ordinance will be taken.

Additional Readings

Aguilera, Roberto F. and Marian Radetzki, "Shale Gas and Oil: Fundamentally Changing Global Energy Markets," *Oil and Gas Journal*, December 2, 2013.

American Society of Civil Engineers, 2013 Infrastructure Report Card.

Berry, Brian and John Parr, *Market Centers and Retail Location Theory and Applications* (Engelwood Cliffs, NJ: Prentice Hall, 1988).

Coupal, Roger H. and David Holland, "Economic Impact of Electric Power Industry Deregulation on the State of Washington: A General Equilibrium Analysis," *Journal of Agriculture and Resource Economics*, 27, 1 (2002): 244–260.

Deno, Kevin T., "The Effect of Public Capital on U.S. Manufacturing Activity: 1970–78," *Southern Economic Journal*, 55 (October 1988): 400–411.

Manyika, James, Michael Chui, Brad Brown, Jacques Bughin, Richard Dobbs, Charles Roxburgh and Angela Hung Byers, "Big Data: The Next Frontier for Innovation, Competition and Productivity," McKinsey Global Institute, May 2011.

McClanahan, Robert, "Electric Deregulation," *IEEE Industry Applications Magazine*, March–April, 2002.

Munnell, Alicia, "How Does Public Infrastructure Affect Regional Economic Performance," *New England Economic Review* (September/October 1990): 11–32.

Keating, Raymond J., "Energy Cost Index 2012: Ranking the States," Small Business and Entrepreneurship Council, June, 2012.

US Energy Information Agency, 2012 Annual Energy Outlook.

Chapter 7

Land Use and Energy Led Economic Development

Chapter Goals

1. Understand how land use policy impacts economic development.
2. Recognize the connection between land use policy and energy led economic development.
3. Be prepared to discuss the role the Millennials play in an energy led economic development strategy.
4. Understand the economic impact of sprawl.
5. Recognize the economic value of compact development patterns and urban redevelopment.

How and where America develops property impacts energy consumption. American land use policies substantially contribute to the expanded consumption of energy, driving up the cost of energy and the cost of doing business in a region and state. America's sprawling land use development pattern is the prime reason for this expanded energy consumption, and the adoption of an energy led economic development strategy from a land use perspective can drive down energy consumption and energy cost in a region and state.

A land use energy led economic development strategy centers around both developing unimproved land in smarter, more densely patterned developments and adopting urban redevelopment strategies to redevelop existing sites. These tactics together will turn fewer cornfields into new homes and strip commercial retail centers. These tactics together will consume less energy and, in some cases, utilize existing traditional and renewable energy sources already in place.

More compact new developments and redeveloping urban centers are not only environmentally conscientious, but also reduce regional infrastructure costs by stemming the tide of urban sprawl. The post-World War II suburban explosion led to a substantial increase in the demand for oil, as the number of car miles grew to connect workers with their suburban homes and central business district jobs. More compact new developments and redeveloping the urban core reduces infrastructure spending and promotes energy consumption by concentrating customers around existing energy infrastructure.

Promoting more compact new developments and urban redevelopment as part of an energy led economic development strategy is gaining steam by capitalizing on a renewed market demand for urban housing. By 2025, the 67,000,000 new Americans added to the population base will need 35,000,000 new housing units and 17,000,000 rebuilt housing units.[1] Recent trends indicate that the housing market is moving more up than out. From 2007 to 2009, building permits for new single-family housing construction have fallen by 42 percent but permits for multi-family units have only dropped by 11 percent.[2] Whatever growth and development in the housing market is occurring involves moving into a more densely populated land use market.

This drive for compact, urban mixed-use developments is responding to the interest of young workers, referred to as the millennial generation. Unlike other urban redevelopment strategies of previous decades, today's urban redevelopment strategies have a greater opportunity to succeed because a large population base of young workers wants to live and work in a mixed-use, urban setting.

Regions and states are adopting many strategies to promote energy led economic development and recruit the millennial generation to their community. The goal is to revitalize the central business district or Downtown of the region into a residential, commercial, and retail center, and start new, compact developments surrounding the existing city's footprint. Tactics to achieve this goal include adopting smart growth land use policies to manage sprawling development; creating land banks to redevelop an existing housing base; redeveloping contaminated Brownfield and historic sites into new, productive uses; building energy districts that operate on their own sources of power; and adopting innovative land use strategies to promote urban redevelopment and compact new developments through Transferred Development Rights and Transit Oriented Development programs. These tactics are producing results and creating energy led economic development strategies that produce young workers interested in energy industry jobs and developments that reduce the overall energy consumption costs of a region.

The Costs of Urban Sprawl

Addressing urban sprawl is a challenge that needs to be addressed through an energy led economic development strategy if a region or state is interested in recruiting energy-oriented companies. The burden urban sprawl puts on taxpayers is large. American local governments spend nearly $140,000,000,000 to build new community infrastructure and over $200,000,000,000 on recurring service costs to keep these new communities running.[3] Sprawl without a doubt contributes to America's substantial use of oil. Oil constitutes over one-third of all American energy consumed, and that large percentage of usage is directly related to American use of the automobile. Oil runs cars and the extension of America's cities into a suburban nation is empowered by Americans driving oil-consuming cars. High sprawl regions drive cars more often—on average 180 cars per 100 households for high sprawl regions compared to 162 cars per 100 households to lower sprawl regions.[4] Sprawling regions have more air pollution, more traffic congestion, and a greater risk of fatal accidents.[5]

Energy Led Economic Development and Urban Redevelopment

America's cities are experiencing a great urban migration. The Baby Boom Generation, followed by their children, ran away to the suburbs from the cities. The American suburban boom played out in all regions of the United States and created a sprawling development pattern fueled by low-cost gasoline and an interstate highway system that eliminated the challenge of distance for many outlying communities. Outer belts built to prevent interstate travelers from being ensnarled in Downtown traffic in reality became major economic hubs. Outer belts circling the city then fueled more new home growth where farm fields once dominated.

The millennial generation is changing American development and energy usage patterns. Millennials are those that are 18–36 years of age and they constitute a large economic demographic. As 10,000 Americans a day reach retirement age, the ability of regions and states to recruit Millennials to their workforce become more and more important. Recruiting Millennials may not be as easy as it sounds. The consumer perspectives of the Millennials create a changing landscape of the energy industry. This changing landscape creates economic development opportunities for regions and states sensitive to the wants and needs of Millennials. Regions and states able to adapt and attract Millennials will generate a large pool of workers for decades to come. Regions and states looking to attract Millennials need an energy led economic development strategy catered to the needs and interests of this younger generation.

The Millennials

The millennial generation in the United States constitutes one-third of the American population and is the largest demographic group in the United States.[6] The largest cohort among the Millennials are 23 year olds; thus, Millennials will be a large population force for some time to come.[7] Those between 18 and 36 years of age are a major economic force. By 2025, Millennials will constitute 75 percent of the America's workforce and already constitute $1,000,000,000 in consumer spending.[8] Millennials are also more highly educated than previous generations, with 47 percent holding a college degree.[9] A major difference between Millennials and post-World War II generations is they are more loyal to a place than to a company. Children of the Great Depression and their children who lived through the economic turmoil of the 1970s often had greater loyalty to a job than to a region or state. Millennials do not have that perspective. The quality of the place to which they are looking to relocate is the first issue they address. They find the place they like, and then they find a job.[10]

Millennials own fewer cars and they are impacting the auto industry and changing American cities. From 2001 to 2009, the average number of automobile miles driven dropped by nearly 25 percent among those 16–24 years of age.[11] From 2006 to 2009, 1.5 percent fewer 16–34-year-olds were driving themselves to work and were much more likely to use public transit.[12] In fact, Americans under 30 used public transit two to three times more often than those older than themselves.[13] The Millennials' car attitude is influencing the overall miles driven. The average American annual car miles traveled is currently at the same levels as 2005.[14]

This new generation of young workers operates with a different perspective on where and how they want to live, and they are changing the urban landscape all across the United States. 62 percent of Millennials want to live in mixed-use developments, 40 percent of Millennials want to live in urban—not suburban—area, and, most surprising, two-thirds of Millennials are renters.[15] Millennials are looking for more of a metropolitan community rather than a suburban community. The Millennials are more inclined to be renters than homeowners. In the first quarter of 2014, the homeownership rate for Americans 35 and under declined to a record low of 36.2 percent.[16]

Millennials also have distinct opinions about energy issues. Millennials are a socially conscientious generation and are expected to make decisions concerning utilities to match their ideals and not just their wallets. Millennials are strong supporters of renewable energy and a sustainable development model.[17] 82 percent of Millennials support additional federal funding for wind, solar, and hydrogen technologies.[18] Millennials additionally support corporate social responsibility and they want to purchase from and work for companies with a social conscience.[19] In the energy context, Millennials are attracted to utility services that develop power from renewable sources. Finally, Millennials are the energy workforce of the present and future.[20] No energy company can succeed without recruiting and retaining young workers. Appealing to the perspective of this young workforce affects the ability to recruit and retain these workers.

Recruiting the Millennials

Regions across the United States are battling to recruit the Millennial Generation. Many strategies for Millennial recruitment exist, but the development of an energy led economic development strategy centering primarily on changes to American mindsets on land use is appealing to America's younger generation. Midsized cities in particular are implementing "Downtown development strategies" that are producing substantial results in the recruitment of residential projects to the region's central business district. Cities across the United States are restoring Downtown parks and green spaces, promoting the integration of commercial and residential uses in multistory former office towers; building public transit lines usually in the form of light-rail systems; redeveloping Downtown infrastructure to support pedestrians over cars; redoing connecting landscaping with public infrastructure; converting surface parking lots into green spaces; promoting the development of educational, arts, and entertainment facilities to Downtown; redeveloping retail centers at traditional locations; and avoiding the 1990s trend of big-box, mega malls Downtown.[21]

The successful urban revival of America's Downtown is happening because of strategic redevelopment strategies that are good not only for the redevelopment of a city's urban core, but this is also a step to stemming the tide of urban sprawl. Traditional urban sprawl prevention tactics center on applying taxes and fees to developments spreading out not up. A new model is emerging capitalizing on market interests of Millennials to live in mixed-use urban settings rather than the suburb, a county away from the region's central business district. This concentration of customers and workers will drive the office, commercial, and industrial market to focus back where a new base of young workers are located.

Look at Kansas City as an example of a region growing through millennial recruitment. Kansas City launched a $5,500,000,000 Downtown initiative geared in large part to building a residential as well as office and entertainment district.[22] The Kansas City initiative includes a planned $100,000,000 streetcar starter line, development of a major entertainment, several mixed-used developments, and Downtown residential projects.[23] The residential population of the Kansas City Central District increased by 50 percent to 20,000 since 2000, and apartment occupancy is above 95 percent.[24] Residents 25–34 make up 25 percent of Downtown Kansas City's population and the residential boom is continuing to grow.[25] The residential projects in Downtown Kansas are only getting bigger, including:

- The One Light $79,000,000 25-story luxury residential tower;
- $90,000,000 Commerce Tower office conversion to a mixed-use development;
- North Point Development's $63,000,000 residential redevelopment of the Art Deco Power & Light Building; and
- A $63,000,000 new building to develop 389 apartments and 12,000 square feet of retail on a site owned by the Port Authority of Kansas City.[26]

Kansas City's Downtown is struggling to attract employers and jobs, but it is quickly replacing that office space with residential units and developing a millennial residential neighborhood in a community that used to close every day at 5 p.m. The story of Kansas City is being replicated all across the United States in mid-sized cities who are seeing a boom in urban infill residential projects in many cases replacing office space. Energy led economic development strategies offer a range of strategies that can capitalize on this new urban growth to build upon it and create high-wage jobs to match the residential development.

Smart Growth and Energy Led Economic Development

Since the US Supreme Court ratified the power of local governments to zone land in the *Euclid* case in the 1920s, America is on the road to urban sprawl. Traditional Euclidian zoning divides land within a city into zones with clear restrictions on uses in the different zones and size and location of building standards that these users are required to meet. Traditional Euclidian zoning created orderly growth based upon a set comprehensive land use plans, with the goal of limiting interaction of incompatible uses and protecting the public health, safety, and welfare.

Traditional Euclidian zoning decisions did not just establish that local governments could establish a land use planning program and require residents to gain approval from the local government for the type of use planned for a parcel of land. *Euclid* also established American zoning would be a separation of uses. Traditional Euclidian zoning creates auto-dependent housing subdivisions and segregated land uses. Each neighborhood in a Euclidian zoning model consists of housing subdivisions, retail shopping centers with ample parking usually out front of the store, office centers, again with ample car parking out front, civic institutions served again by their own parking, and roads connecting all of these separated land uses.[27]

In a nation in the middle of the Industrial Revolution, a separation of uses made sense. Public health requirements demanded separating the factory from the school or home. Thus, local governments starting in the 1920s enacted Euclid style zoning ordinances all across the nation. However, an outgrowth of traditional Euclidian zoning is the invention of urban sprawl. As the 1920s gave way to the post-World War II boom and the interstate highway system made rural land outside the cities easy to access with a car, the suburbs became cheap and easy to develop. Developing unimproved land is cheaper than developing improved land because improved land has higher transaction costs.[28] Zoning adds to this transaction cost for improved land through the regulatory process. Rural communities on the outskirts of urban centers lacked the complex zoning regulations and processing hindering their urban neighbors. Mix in federal subsidies for home buyers and the American Dream of owning a home, and the result is urban sprawl.

Economic and Energy Efficiency Advantages of Compact Development Patterns

Developing at higher densities uses land more efficiently and generates greater economic return. Sarasota County, Florida, generates more property tax revenues per acre than single-use developments in more suburban locations through compact, mixed-use developments.[29] Density mandates do drive up costs, but they can produce substantial returns. In addition, higher density, mixed-use development can reduce the public infrastructure costs of providing public infrastructure and delivering services.[30]

Greater density in development can also save billions in future taxes to cover the cost of additional infrastructure to meet the demand for new growth. A study of the Salt Lake City region showed that the adoption of a more compact land use development pattern could save $4,500,000,000 over 20 years through a new growth strategy impacting transportation investments, and zoning,[31] In addition, compact development patterns can actually produce lower costs for developers as well as the government. Compact development patterns create infrastructure efficiencies requiring shorter utility, road, water, and sewer runs. Infrastructure improvements are a major cost source for developers, and more compact development patters with residential and commercial districts on small lots connected closer together reduce infrastructure costs compared to traditional, large residential lots connected to even larger commercial districts by a complex web of connecting roads seen in suburbs across the United States.[32]

Density in land use patterns also promotes a more efficient energy system. The efficiency of energy consumption in the service sector is higher for densely populated cities, and energy efficiency increases by 12 percent when a city's population density increases.[33] Land use density impacts energy consumption related to transportation energy. The greater the population density, the less oil is consumed, as distances traveled are lessened and the use of mass transit becomes more economical.[34] Sprawling cities such as Houston and Phoenix top the major city list for transportation fuel consumption, while highly dense New York City consumes the least amount of fuel on a per capita basis among major American cities.[35]

Ramapo and Emergence of Smart Growth

Sprawl is the real estate development process in which the spread of property improvements far outpaces population growth. Real estate development reaches the sprawl phases when population is dispersed in low-density developments with rigidly separated homes, shops, and workplaces. These areas are served by a network of roads marked by huge blocks and poor access and a lack of well-defined, thriving activity centers.[36] A lack of compact development patterns provides a negative fiscal impact on government. America's sprawling development pattern costs big bucks. Fueled not just by Euclidian zoning, but also by a trillion dollar interstate highway system, America's surface transportation infrastructure needs $1,700,000,000,000 before 2020 in highway and transit projects.[37]

Smart growth, just like *Euclidian* Zoning, is guided by the United States Supreme Court as to how development can be managed without violating private property rights. In 1972, the US Supreme Court dismissed the appeal of a New York case, *Golden v. Planning Board of the Town of Ramapo*.[38] In *Ramapo*, the local government adopted a land use management plan by establishing an 18-year capital improvement program that linked development with the availability of public facilities. Since the court's support for this model, smart growth strategies have centered on what are really Ramapo plans in a legal sense.

Smart growth is a land use planning strategy that addresses sprawl and is part of an energy led economic development strategy. As table 7.1 illustrates, smart growth uses comprehensive land use planning to guide, design, develop, revitalize, and build communities to promote greater density in development that reduces energy consumption, promotes mass transit, encourages mixed-use development, and is pedestrian-oriented.[39] Smart growth furthermore focuses growth along transit corridors, connects housing closely with job centers, increases the density and reduces overall lot sizes, and ensures that adequate public open spaces are provided. Smart growth is impacted by state laws but most often implemented at the local government level.

Successful smart growth strategies are a two-pronged approach geared toward redevelopment of the urban core as well as the management of suburban growth into more compact, environmentally friendly developments. Both of these approaches create a development pattern that is appealing to a large new energy workforce but also results in lower energy consumption. Lower energy consumption, through the provision of power for homes and businesses and less transportation fuel consumption, make a region and state more appealing from a cost standpoint. Too often, the term smart growth sends public and private sector players right to the thought of new taxes and impact fees. Smart growth can involve new taxes and impact fees, but communities that see smart growth as a new revenue source are missing the true energy led economic development benefits. The two-pronged smart growth strategy focusing on both redeveloping the urban core and

Table 7.1 Smart growth tactics

Transferred Development Rights
Transit Oriented Development
Regional Planning
Energy Zones
Land Banks
Historic Preservation
Brownfield Redevelopment

creating more compact, suburban developments will consume less energy and act as an economic stimulus to the region.

Transferable Development Rights

A Transferable Development Right (TDR) is a smart growth land use regulation strategy in which the development potential of specified land is traded on the open market for other specified land.[40] Land is said to be protected from development that either is environmentally sensitive or if there exists an artificial boundary line, where sprawl is stopped. A TDR acknowledges that owners of what would otherwise be sprawling property need to be compensated in some way to prevent an economic injustice as well as to prevent sprawl from occurring. TDRs preserve historic structures and protect agricultural lands, forestlands, and open space, and the ultimate goal is to shift development in rather than out.

Sprawling suburban Washington, DC is an example of a region looking to TDRs to manage growth. Montgomery County, Maryland enacted a TDR that downzoned rural areas from one dwelling unit for every 5 acres to one dwelling unit for every 25 acres.[41] A permanent easement to regulate residential density was created and landowners were given five transferable rights, allowing them to build one additional dwelling unit per each 25 acres.[42] These development rights can be purchased.[43] In 2000, the market value for a single TDR right was approximately $10,000, and more than 38,000 acres of the approximately 91,000 rural acres had been preserved.[44]

Energy Zones

Regions looking to more directly link land use and energy consumption are creating energy districts. Energy districts are defined geographic regions where the energy consumed is created by a range of means within the district itself. Energy from outside sources is not used, and a batch of subsidies and incentives are associated with the energy district to ensure that the cost of energy does not prevent economic development.

Fort Collins, Colorado's Zero Energy District, offers an example of a model for an energy district that promotes energy efficiency and economic development. The Fort Collins Zero Energy District capitalized on the local municipal electric company, five companies open to this novel energy approach, and a number of local renewable energy companies interested in a national energy district model.[45] First, the buildings within the Fort Collins Zero Energy District were modified to become as energy efficient as possible, and they reduced their energy consumption by 20 percent during peak periods.[46] Smart meters were installed to promote energy efficiency.[47] This successful demonstration project required energy efficiency investments in power generation, storage, efficiency, control, communication equipment, technologies, and strategies.[48] Fort Collins utilized sited solar photovoltaic arrays, and companies within the district used new electricity generators powered by natural gas, biogas, fuel cells, thermal devices, and energy-storage devices.[49] Roughly $10,000,000 in federal and utility funding was used to develop the project.[50]

Property-Assessed Clean Energy (PACE) financing is a method companies can use to fund energy district initiatives. This public finance tool is authorized by state law and utilizes planned energy savings to gain bond financing to fund energy efficiency measures. New Hampshire authorizes its cities, towns, and village districts to establish energy efficiency and clean energy districts that can incur debt through a bond fund, establish revolving funds, provide financing, and collect assessments to implement a PACE program.[51] $5,000 is the minimum total assessment for a single-family property and $35,000 or 15 percent of the assessed value is the maximum for the property multiplied by the municipality's current equalization ratio, whichever is less.[52] For other properties, $60,000 or 15 percent of the assessed value is the maximum.[53]

Regional Planning: Urban Growth Corridors

Booming regions are natural locations for urban sprawl. Rising property values and a large transportation infrastructure network connecting urban, suburban, and rural communities drive growth and development to more affordable rural land further from the urban core. Left unregulated, farmland becomes sprawling housing divisions with retail and commercial centers soon to follow.

States such as Oregon have put a stop to this natural economic expansion by concentrating their growth and development into a defined geographic district. Oregon is a haven for those in Northern California looking to leave the growth and chaos of that region. Oregon's 1950 population hovered around 1,500,000 and now is approaching 4,000,000.[54] Oregon is adding population at an aggressive rate.

Oregon uses a state and local comprehensive land use plan consistent with statewide planning goals.[55] The Oregon Land Conservation and

Development Commission reviews comprehensive plans proposed by state, county, city, and other local agencies for consistency and compliance with statewide goals and laws.[56] Oregon's cities and metropolitan areas create an urban growth boundary around their perimeters, which operates as a land use planning line to control expansion. Portland Metro manages the region's urban growth boundary. The Portland Metro is a governmental body that performs regional governmental operations and planning functions.[57] Portland Metro creates regional land use goals and policies on land use, transportation, parks and open spaces, water management, and natural hazards for its urban growth boundary.[58] Land inside the Portland urban growth boundary is prioritized for public infrastructure investment and development outside the boundary is not permitted beyond its current use. Every six years, the urban growth corridor is reviewed, a 20-year population forecast is provided, and the boundary of the corridor can be adjusted.[59] Portland's area urban growth boundary expanded three dozen times since first drawn in 2007.[60]

The economic success of Portland is well known. The region continues to grow and develop and it proves that growing regions can manage that growth and still provide high-wage jobs for its residents. Such aggressive growth management tools might not work for every state and region in the United States, but they clearly have in Portland.

Transit Oriented Development

Transit-Oriented Development (TOD) is another smart growth strategy designed to create compact developments and spur redevelopment of the core city. A TOD utilizes zoning and incentives to spur development within walking distance of a transit stop. TODs will mandate mixed-use developments, create walkable and pedestrian friendly spaces, and reduce parking mandates to ensure the development does not look like one giant parking lot.

TODs increase density along transit corridors by promoting the location of housing and job centers and retaining shops connected to transit stops. TODs create this connection by promoting mixed-use developments as close to the transit stop as possible. In major urban areas, with a young population looking to avoid owning a car, living and working near a transit stop matters. TODs, besides serving as a recruiting tool for Millennials, also lower traffic congestion, improve air quality and reduce overall fuel energy consumption by promoting the use of mass transit and increasing the population density in urban markets.[61] In fact, residents in transit rich communities spend 16 percent less on transportation costs.[62]

Arlington Heights, Illinois, a Chicago suburb, offers an example of economic success using TOD. Utilizing a TOD framework, Arlington Heights turned a train commuter center into their new town center by creating a mixed-use development with a performing arts center, high-density housing, and commercial centers and well placed public parking garages.[63] The results of the Arlington Heights TOD are impressive. From 1980 to 2000, residents living in Downtown Arlington Heights jumped from 350 to 2,200, and, since 1997, $27,000,000 in public investment has leveraged $225,000,000 in private investment.[64] A major incentive for this Downtown Arlington Heights boom is the construction of a new commuter rail center in this neighborhood.

Land Banking Program

An effective means to prevent sprawl is to redevelop underutilized land within an urban setting. America's urban centers have no shortage of vacant, tax delinquent, and abandoned housing and commercial centers. Sprawl can be impacted by developing these existing housing and commercial sites that already have infrastructure in place through transit, roads, highway, bridges, power, and a large population base to provide workers and customers. Land banking is one tactic used by local governments to redevelop existing neighborhoods where the government purchases land to preserve as green space or redevelop for housing or commercial purposes. The government purchases vacant, tax delinquent, and abandoned housing and commercial centers, clears or remodels the structures on the property, and then gives it away or resells it at a subsidized rate to a new owner.

One of the biggest challenges with urban sprawl is that it is not limited to growing and economically successful areas. Sprawl happens in struggling regions just as often as it does in growing cities. Look at Michigan for just such an example. Flint, Michigan is one of many communities living thorough the "boom and bust" cycle on a single industry—in this case, the auto industry. The site of a major General Motors facility, Flint's population peaked in the 1960s to around 200,000 people but, like Detroit, as the American auto industry struggled, so too did Flint. Now its population is nearly half its 1960s high. With the loss of population comes an explosion of vacant, tax delinquent, and abandoned property overtaking neighborhoods like a cancer spreads. Flint's answer in part to their current economic struggles is the creation of a land-banking program to redevelop the infrastructure and community.

State law permitted the creation of a land bank operated by the county government to purchase and develop this vacant, tax delinquent, or abandoned property. The Genesee County Land Bank Program serving Flint operates nearly 9,000 sites—including 146 commercial buildings 4,477 residential homes, 206 commercial vacant lots and 4,108 residential vacant lots.[65] Michigan state policymakers assisted in the development and success of Flint's land bank program by revising and expediting the state's land foreclosure process. Michigan County Treasurers were given outright ownership of tax delinquent properties after two years.[66] The previous law took four to seven years for ownership transfer to revert to the local government and resulted in the decline of once-great Michigan cities.[67] Flint, Michigan and other Michigan communities now have more control over underutilized properties as a prime redevelopment tool.

As figure 7.1 illustrates, speeding up the foreclosure process is only part of the reason for the success of the Genesee County Land Bank program. Michigan also creates funding mechanisms for local land bank programs. Michigan allows land banks to recapture 50 percent of the property tax revenues for the first five years after transfer of property to a private party.[68] This provides an on-going revenue stream for the land bank. Michigan land banks are also permitted to borrow money, issue tax-exempt financing, and select properties to acquire from tax delinquency roles.[69]

As a result of Michigan' land bank program, Flint is putting property to actual productive use.

700 Flint lots have been transferred for side yards, 90 affordable rentals and 80 single-family homes have been reconstructed, and the land bank's $3,500,000 demolition program over three years increased surrounding property values by more than $100,000,000.[70]

- Narrow focus with clear goals and objectives for vacant land reutilization;
- Close coordination among city departments and external partners;
- Expedited judicial foreclosure process
- Independently established land banks with a corporate structure
- Integrated IT management system containing parcel-specific information;
- City-wide strategic vision integrated with land bank planning;
- Streamlined eminent domain process;
- Ability to determine the terms and conditions for sale of properties; and
- Diverse funding streams

Figure 7.1 Keys to successful land banks.

Source: "Best Practices in Land Bank Operation, Cleveland State University, June, 2005.

Brownfield Redevelopment

Brownfields are real properties contaminated by hazardous substances such as petroleum products.[71] However, Brownfield programs exclude facilities listed on the US EPA Superfund List, facilities subject to unilateral administrative orders, and government-owned property.[72] Federal and state environmental laws create liability for contamination at sites that freeze development. Federal environmental law creates liability for site contamination for anyone that owns the property, whether they polluted it or not. However, a "bona fide prospective purchaser defense" can be established before purchasing contaminated property if the buyer establishes that a third party is responsible for the pollution. The United States' 500,000–1,000,000 Brownfields are an economic development opportunity waiting to happen as they often have infrastructure in place and are surrounded by a strong workforce and customer base.

Numerous state and federal Brownfield programs provide environmental protection and funding to clean up the sites and redevelop the land. Federal Brownfield funding is primarily used to determine the extent of the environmental damage and create a remedy plan to permit the land to be used again. The US EPA offers funding for Brownfield assessment, cleanup, revolving loans, and environmental job training. Assessment grants provide $200,000–$700,000 to fund the inventory of sites, assess the contamination, and develop a remedy to clean up the site.[73] Federal cleanup grants provide only $200,000 per site and the applicant needs a 20 percent cost share. Unfortunately, many Brownfield sites require far more than $200,000 to remedy their contamination. Job training grants of $300,000 for up to five years for Brownfield sites are also provided to nonprofits, colleges and universities, and local and state governments.[74] $1,000,000 over five years can be gained by local and state governments from the Brownfield Revolving Loan Fund as well.[75] These federal government Brownfield programs leverage $17.39 for every EPA dollar spent and created over 70,000 jobs.[76]

With federal funding not really close to large enough to address the number of Brownfield sites, state governments also are working to address this economic opportunity. Over 10,000 Brownfields were remediated recently by states, municipalities, and tribal agencies and these Brownfields Programs leveraged more than $14,000,000,000 in other financing, creating over 60,000 jobs.[77]

Florida offers just one example of a state Brownfield redevelopment program. First, Florida law provides a mechanism by which property owners who are not responsible for environmental contamination can gain immunity for the environmental damage.[78] Florida provides a Voluntary Cleanup Tax Credit to address contamination at dry cleaning and other sites in designated brownfield areas, which constitute two-thirds of Florida's 67 counties.[79] Public or private companies are eligible for up to $5,000,000 annually in credits against the Florida Corporate Income Tax.[80] Florida also offers a 20 percent tax refund tied to the creation of new jobs at a Brownfield site. A $2,500 cap per job is provided with no more than 25 percent of the total refund approved used in any single fiscal year.[81] Florida's Brownfield program offers many success stories. The City of Pompano identified a contaminated site and received $527,000 in voluntary tax credits.[82] Based upon the clean-up plan and tax credits, this once polluted site will be the location for a redeveloped neighborhood, with a 46,000 square-foot library and cultural center adjacent to Pompano Beach City Hall.[83]

Historic Preservation

Saving historic structures can be a key element to creating more compact developments and redeveloping the urban core. Historic preservation increases land values and impacts economic development. [84] Historic Preservation projects created more than 2,000,000 jobs and over $90,000,000,000 in investment.[85] Rehab construction creates 50 percent more jobs than new build construction.[86] Historic preservation revives neighborhoods, promotes the use of existing infrastructure, and reduces energy consumption by drawing residents and businesses back to the core of the city.[87] Historic preservation projects save big on infrastructure costs—50–80 percent, compared to new suburban development.[88]

Historic Preservation is a government designation of a structure as well as an array of tax incentives that make the redevelopment of a historic structure attractive. Historic preservation uses local zoning regulation to determine whether any changes to buildings can be made, as well as the style of changes that can be made to a structure if the building itself is historic in nature or the neighborhood is historic in nature.[89]

Along with local government regulation related to the use, remodeling, or destruction of historic structures comes an array of state and federal government incentive programs to fund the redevelopment of historic structures. The Federal Historic Tax Credit supports 1,000 projects annually with a 20 percent tax credit for historic properties, a 10 percent tax credit for non-historic properties built prior to 1936, and tax benefits linked to historic preservation easements.[90] One limitation on the federal historic tax credit is that the facility needs to be income producing, which excludes government facilities and other public-private partnerships. The Federal Historic Tax Credit not only spurs urban redevelopment, but also produces a net gain in tax revenue for the federal government, as once-dormant property becomes productive property.[91]

States across the nation operate state-based historic preservation programs to encourage the redevelopment of historic structures. As an example, North Carolina offers a 20 percent tax credit for those taxpayers who receive the federal historic preservation credit.[92] This provides a combined 40 percent credit and doubles the economic incentive for historic projects. North Carolina also offers a 30 percent credit for the rehabilitation of non-income-producing historic properties that includes private homes.[93] Finally, North Carolina offers State Mill Rehabilitation Tax Credits to redevelop former industrial sites.[94] Since 1976, North Carolina reviewed over 2,000 completed "certified rehabilitation" projects, creating over $1,000,000,000 in economic investment.[95]

Big Questions

1. How do struggling communities adopt "smart growth" without killing any potential development?
2. Can the Portland model work in other places?
3. Why should government subsidize urban redevelopment when the Millennials create such a strong market for this land?
4. How is urban redevelopment related to energy led economic development?
5. Will land banking programs really turn the tide in struggling cities like Flint, Michigan?

Class Writing Assignment

The Mayor of Las Vegas is promoting the concept of an urban boundary zone for the southern Nevada region. The battle will be fierce, as the advocates for continued growth and sprawl are strong. The Mayor of Las Vegas asked you to draft a one-page policy brief that can be used in advocacy efforts to outline the challenge of sprawl in Las Vegas, how compact developments and urban redevelopment are central to attracting the all-important Millennial workforce, how an urban boundary zone would work in Las Vegas, and why a Las Vegas urban boundary zone would promote energy led economic development.

Class Debates

Remember, the Mayor of Las Vegas is pushing her smart growth plan to create an urban boundary around existing suburban Las Vegas to stem the sprawl from around the region. The only challenge is that she needs the support of the Republican-dominated General Assembly. The Governor is on board. State Senator Huey Long of Las Vegas introduced legislation mandating the creation of an urban boundary zone around Las Vegas modeled on the Oregon law. Legislative debate is happening. The state's developers are organized and ready to oppose the measure and have lined up the Nevada Chamber of Commerce. Supporters of the bill include the City of Las Vegas, all its suburbs, and MGM and the other major casinos in the state. Senator Long and his staff need to draft the legislation, prepare, and give testimony before the Nevada Senate Jobs and Economic Development Committee. Members of the Committee need to listen to all the testimony given from proponents, opponents, and sponsors of the bill, debate the legislation, ask questions, and then vote on the bill.

Additional Readings

Alexander, E. R., "A Transaction-Cost Theory of Land Use Planning and Development Control: Towards the Institutional Analysis of Public Planning," *The Town Planning Review*, 72, 1 (January 2001): 45–75.

Berton, Brad, "Giving as Good as They Get: Zero-Energy Districts," Urban Land, February 21, 2013, retrieved from http://urbanland.uli.org/sustainability/giving-as-good-as-they-get-zero-energy-districts/.

Calthorpe, Peter, *The Next American Metropolis: Ecology, Community, and the American Dream* (New York: Princeton Architectural Press, 1993).

Duany, Andres, Elizabeth Plater Zyberk, and Jeff Speck, *Suburban Nation: The Rise of Sprawl and the Decline of the American Dream* (Union Square West, NY: North Point Press, 2000).

Ewing, Reid, Rolf Pendall, and Don Chen, *Measuring Sprawl and Its Impact* (Washington, DC: Rutgers University, Smart Growth America, 2002).

Frellich, Robert and Neil Popowitz, "The Umbrella of Sustainability: Smart Growth, New Urbanism, Renewable Energy and Green Development in the 21st Century," *Urban Lawyer*, 42, 1 (Winter 2010): 1–39.

Gilderbloom, John I., Matthew J. Hanka, and Joshua D. Ambrosius, "Historic Preservation's Impact on Job Creation, Property Values and Environmental Stability," *Journal of Urbanism: International Research on Placemaking and Urban Sustainability*, 2, 2 (July 23, 2009): 83–101.

Gose, Joe, "Millennials Going to Kansas City, to Live and Work." *New York Times*, August 19, 2014.

Katz, Peter, "Sarasota's Smart Growth Dividend." Planning. American Planning Association, December 2010.

Kemp, Roger L., "The American Dream Moves Downtown," retrieved from http://www.wfs.org/content/american-dream-moves-downtown.

Morikawa, Masayuki, "Population Density and Efficiency in Energy Consumption: An Empirical Analysis of Service Establishments," *Energy Economics*, 34, 5 (September 2012): 1617–1622.

Muro, Mark and Robert Puente, *Investing in a Better Future: A Review of the Fiscal and Competitive Advantages of Smarter Growth Development Patterns* (Washington, DC: The Brookings Institution, 2004).

Newman, Peter and Jeffrey Kenworthy, "Urban Design to Reduce Automobile Dependence," *Opolis: An International Journal of Suburban and Metropolitan Studies*, 2, 1 (2006): 35–52, Article 3.

Speir, Cameron and Kurt Stephenson, "Does Sprawl Cost Us All? Isolating the Effects of Housing Patterns on Public Water and Sewer Costs," *Journal of the American Planning Association*, 68, 1 (2002): 56–70.

Chapter 8

Building an Urban Energy Center

Chapter Goals

1. Understand the economic benefits of building an urban energy center.
2. Understand the economic measures that make Houston a global energy center.
3. Be prepared to discuss the economic history of Houston.
4. Recognize how Houston became a global energy center.
5. Understand what role the state of Texas and federal government play in making Houston a global energy center.

The energy industry is the basis for many successful urban centers. The larger question is how a city, region, and state can capitalize on the discovery or development of renewable or nonrenewable energy sources to build a successful regional or even global economic success story. Houston, Texas offers a clear model for how a city can use its energy resources and turn itself into a global energy powerhouse. More impressively, Houston offers a model for how the growth and development of the energy industry can be used to diversify a regional economy and produce long-term economic success. Just look at the impact of ExxonMobil. ExxonMobil's Woodlands, Texas new corporate campus will house 10,000 employees about 25 miles north of Downtown Houston. ExxonMobil's Woodlands Campus spurred growth in the healthcare industry following this massive new development.[1] CHI St. Luke's Health announced an $110,000,000 outpatient medical complex near ExxonMobil's new campus to complement its hospital nearby, and Houston Methodist Hospital announced a new 193-bed, $328,000,000 hospital in The Woodlands.[2]

Houston Is a Global Economic Powerhouse

To say Houston is an economic success story is a massive understatement. Houston is a megacity and is the fourth largest city in the United States. Houston is a job-producing machine with a low cost of living and diversified economic base. Houston has a population of 6,100,000 in the metro region, which recently added 313,800 new residents, while New York, Los Angeles, and Chicago lost population.[3] Houston is the world's fortieth largest economy with a GDP of $399,700,000,000, and is continuing to grow, with a recent employment increase of 3.3 percent.[4]

The Houston metro center leads the nation with a 4 percent employment increase from July 2013 to July 2014 among the 20 largest metro regions.[5] This growth is not new. This region grew by more than 4 percent in 1997–1998, 2006–2007, and 2012.[6] Since the low point of the Great Recession in 2010, Houston created more than 420,000 jobs (constituting a 17 percent increase) with only New York creating more jobs in that same timeframe.[7] Houston is growing in all industry sectors with 18,900 jobs in the professional, scientific, and technical services area, 15,100 health care and social assistance jobs, 12,400 food services jobs, and double digit percentage gains in building construction and engineering services.[8]

Houston is also an export leader. In 2013, Houston led the nation for the second consecutive year in exports, shipping nearly $115,000,000,000 in goods overseas.[9] This 4.2 percent increase over 2012 gave Houston a slight lead over New York and well ahead of third-placed Los Angeles.[10] Houston's exports have grown by over $73,000,000,000 over the last two years, which is larger than any other metro region.[11]

Houston also recovered from the Great Recession to a larger degree than did any other major metro region. Houston recovered 178.6 percent of the jobs they lost during the more recent recession.[12] Building permits are booming in Houston, with construction permits hitting a record $7,600,000,000 which constitutes a 30.5 percent increase over the year ending in July of 2013.[13] Inflation is also in check in Houston. The regional consumer price index grew at a 2 percent rate nationally from July 2013 to July 2014, and Houston's inflation grew at a similar rate at 1.9 percent.[14]

Home sales in Houston grew by 7.1 percent from 2013 to 2014 and Houston's average home value of $124,700 makes it affordable to purchase a house, as the national home value is over $180,000.[15] This low cost of living is extended to the cost of labor, primarily because only 3.4 percent of Houston's labor force is unionized.[16] Houston is the nation's fifth largest office market and ranks seventh in energy-efficient buildings.[17] Houston's Purchasing Management Index, an economic indicator for regional production, moved from 56.4 in July 2014 to 52.4 in July 2013.[18]

The diversity of Houston is impressive. Houston is a young city with a median age of 33.4 years, well under the US median age of 37.3. Houston's median age is lower than Phoenix, Atlanta, Los Angeles, Chicago, Washington, DC, New York, Philadelphia, and Boston. Of the 2,100,000 people living in Houston, more than 1 in 5 is foreign born, with the nation's fifth largest Hispanic population. In fact, Houston's Hispanic population is larger than the total population of 15 states and the District of Columbia.

Houston's economy is not perfect. Houston only meets the national average for the percentage of its residents with bachelor's degrees, as well as the mean travel time to work.[19] Of greater concern is Houston's poverty rate of 22.2 percent, which is well above the national average of 14.9 percent, and their home-ownership rate of 45.9 percent is 20 points below the national average.[20] Another area of concern is Houston's per capita income of $27,029, which, though higher than the average in the state of Texas, is just below the national average of $28,051.[21] Houston is also massive, covering nearly 600 square miles, which may at some point cause the delivery of services to struggle to keep up with the geographic scale of the city.[22]

Even with these negatives, Houston is a global economic power. Houston experienced increases of nearly 40 percent in the number of college-educated workers, remains in the top ten for fastest growing cities, enjoyed a 20 percent growth in jobs, and is a national leader in high-wage STEM jobs all through the last decade.[23] Houston's low cost of living also makes it a top region for consumer purchasing power, and added more multitenant buildings than even New York did.[24]

Houston Is a Global Energy Center

Houston's economic success is based in large part on the energy industry, but other service and manufacturing jobs support the success of this region. Like nearly all American cities, the service industry leads the Houston economy when Gross Area Product is measured. However, of critical importance to Houston is the fact that mining, which includes the oil and gas industry, is a close second to service, followed by manufacturing in third.[25] Finance, insurance, and real estate tie in a distant third place (figure 8.1).[26]

Houston's largest private sector employers center on energy, healthcare, and service-related companies; energy still accounts for 50 percent of Houston's export-related jobs, but that total is down from 87 percent in the 1980s.[27] Exxon Mobil tops the company list with 21,500 employees in Houston, followed by Memorial Hermann Healthcare System and the University of Texas MD Anderson Cancer Center.[28] In fact, energy and health care constitute half of the top 20 largest employers in Houston,

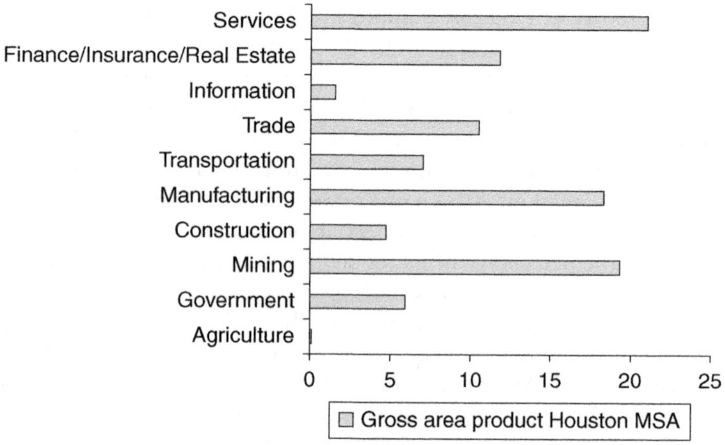

Figure 8.1 Gross area product by industry sector Houston MSA.
Source: Houston facts, Greater Houston Partnership, 2014, p. 14.

and the Texas Medical Center employs 100,000 people and operates a $14,000,000,000 research and development center.[29]

Houston's over-5,000 manufacturers and the region's percentage of manufacturing jobs nearly doubles the national average.[30] As home to the National Aeronautics Space Administration's (NASA) $1,500,000,000 Johnson Space Center, Houston is also home to over 150 aviation and aerospace-related companies drawn to the NASA center.[31] Houston is also a gateway for global trade in the expanding Latin American market, as a transportation and logistics hub for these growing markets. High-tech is also a Houston strength.[32] 190 biotechnology and life science companies call Houston home and the Anderson Medical Center is considered the largest hospital in the United States.[33] Last but not least, 3,700 energy related companies reside in Houston and employ nearly 80,000 engineering and architectural professionals.[34] This large base of engineers positions Houston as a global leader in the development of STEM related jobs.

How Did Houston Become a Global Energy Center?

Houston did not become a global energy leader by accident and it did not happen overnight. Houston also did not become a global energy center simply because of their strategic location near a major oil reserve, nor did this happen because private sector companies just decided they would base themselves in Houston. Houston became a global energy center because of wise and aggressive local business and political leadership capitalized on an oil reserve, who then utilized their political contacts to develop infrastructure critical to the region's recruitment of the oil industry. Once Houston had a foothold in the energy industry, they capitalized on business-friendly state policies in Texas related to annexation and tax policy, but also adopted business-friendly zoning and land use policies and worked to make sure companies had the critical infrastructure and workforce needed to succeed.

A History of Houston and Energy

Houston was founded in 1836 and its location was not an asset. Its inland location was swampy, with no direct route to the Gulf of Mexico. During the 1800s, Houston at best was a center for lumber and distribution, and the most successful Texan city in the region was Galveston, with its large bay ready for transportation and distribution.[35] Natural events and an oil discovery changed the fortunes of these two Texan cities. In 1900, Galveston was crushed by a massive hurricane that killed 6,000 people and critically wounded Galveston Bay.[36] In 1901, large oil reserves were discovered at Spindletop just east of Houston and this discovery propelled Houston into the energy business at just the time that oil was critical to manufacturing production and the operation of the automobile.[37] Oil at the time was becoming the energy of choice, not just for gasoline that operated the newly mass-produced automobile, but also as a home-heating oil.[38] Advancements in drilling technology made oil and natural gas not only abundant, but also available at a cost that could compete with the dominant energy source of the time—coal.[39] Oil and natural gas grew in demand because they were cleaner fuels than coal was, and easier to transport and store for their multiple uses.[40] As the early 1900s turned into the mid-to-late 1900s, American oil consumption increased even more. The explosion of the automobile as the main form of transportation, completion of the interstate highway system, and availability of cheap oil, drove up the usage of oil as America's main energy source.

The Houston, Texas region benefitted greatly from the dominance of oil as America's main energy source starting in the early 1900s. Oil was, fortunately for Houston, the energy source that was discovered in great abundance in their backyard. The Gladys City Oil, Gas, and Manufacturing Company, formed in August 1892, drilled three shallow holes unsuccessfully on Spindletop Hill in 1893 and future efforts over the next several years by others proved equally unsuccessful in this area covered by difficult to drill salt domes.[41] Backed by financial interests in Pittsburgh led by the Mellon Bank, a new heavier and more efficient rotary type bit was used and big results were produced.[42] On January 10, 1901, oil started to flow from Spindletop when drilling 1,139 feet below the surface; oil shot out of the ground to well over 100 feet, producing 100,000 barrels a day.[43] This was an unprecedented oil gusher and, by September 1901, at least six successful wells were drilled in this Beaumont, Texas town.[44] Land speculation went rampant, with land selling for $20,000 that previously had been on the market for $150.[45] Overproduction slowed the oil flow from Spindletop to 10,000 barrels a day in 1904, after producing 17,500,000 barrels of oil in 1902.[46] However, the region remained a mainstay for domestic oil production and put Texas on the map as the new oil capital of America.[47] In fact, over 153,000,000 barrels of oil flowed from Spindletop fields by 1985. This discovery of substantial oil supplies in Texas led also to the development of further oil fields, storage facilities, pipelines, and major refining units in the Beaumont, Port Arthur, Sabine Pass, and Orange areas around Spindletop, and led to the location of major oil company facilities in this region—500 Texas companies were doing business in Beaumont by 1902.[48]

Spindletop's wealth eventually spilled over into the larger neighborhood of Houston. The growth of the oil industry created substantial wealth and a new business elite in Houston. This business elite was determined to further capitalize on the region's success to cement Houston's role as the economic leader in the state of Texas and a global energy giant. A group of Houston business leaders would regularly meet in Suite 8F of the Lamar Hotel to map out regional economic development strategies.[49] The 8F group included industrialists George and Herman Brown; banker, developer, and owner of the Houston Chronicle newspaper Jesse Jones; lawyer Judge James Elkins; insurance entrepreneur Gus Wortham; and oilman James Abercrombie.[50] The 8F group and their predecessors realized Houston lacked the infrastructure needed to succeed in a global economy and that pro-business tax policy and funding was needed from Washington, DC for Houston to succeed.

First, the 8F group figured out what an earmark was before the Congress did. Houston needed access to the Gulf of Mexico and could not get it without financial help from the federal government. The Houston 8F group successfully gained millions in funding that led to the development of the Houston Ship Channel starting in the early 1900s and continuing on to this day.[51] With the costs split between the federal, state, and local governments, Houston built a connection to the Gulf of Mexico that today is the leading port in the United States. The earth did not provide them with this logistics connection; taxpayer and private money built it. With the help of the Houston Ship Channel and continued growth of the oil industry, Houston became the largest city in Texas by 1930.[52] By 1935, 24 percent of Texas oil reserves were within 100 miles of Houston and 40 percent of Texas oil was shipped through the Houston Ship Channel.[53] Continued improvements in the Houston Ship Channel were funded regularly throughout the 1900s to ensure that it would be an attractive location for the distribution of oil to the world. From 1900 to 1963, $63,700,000 in improvements in the Houston Ship Channel were funded, with federal taxpayers covering $60,900,000 and state and local taxpayers covering the rest.[54]

The 8F Group also understood the economic link between manufacturing and distribution. Regional logistics strategies reduce domestic manufacturers' costs of doing business by developing a local supply chain network and building an infrastructure network to sell goods and services beyond the immediate service territory. Logistics are activities that transform and distribute goods, from raw materials to the final product, and then deliver it to global customers.[55] Logistics adds value and is attractive to manufacturers and retailers choosing from among various distribution outlets.[56] The Houston Ship Channel was one of the early logistics plays for a region born without an ocean.

The Houston Ship Channel was not the end of Houston's efforts to utilize the federal government for its economic benefit. As the Great Depression hit in the 1930s, Houston, like most cities, benefited from substantial public works projects that kept people employed and the economy working. As World War II began, one of their own, Jesse Jones, became President Franklin Roosevelt's Secretary of Commerce. Jesse Jones did not forget where he came from. When the nation faced a critical shortage of rubber during World War II and had to develop synthetic rubber, the petrochemical industry was established by the federal government in Houston, with $700,000,000 invested in a defense department facility in Houston.[57] Also, the growth of the defense budget to fight World War II helped Houston, like it did many industrial cities; but Houston was ranked sixth in the nation for wartime industrial plant investments.[58] During the 1940s, $140,000,000 in federal subsidies in two major Texas pipelines ensured Texas oil flowed to New York and Pennsylvania.[59]

President Dwight Eisenhower adopted a federal government policy to limit the import of foreign oil, which benefited large domestic oil producers and the regions they came from, like Houston.[60] The Congress, with the support of Texas' House Speaker Sam Rayburn and Senate Majority Leader Lyndon Johnson, not only secured millions of dollars in federal infrastructure funding for Houston, but adopted an oil depletion allowance that permitted wealthy oil producers to exempt 27.5 percent of their income

Table 8.1 Houston's population growth

Year	Population
1900	44,633
1910	78,800
1920	138,276
1930	292,352
1940	384,514
1950	596,163
1960	938,219
1970	1,233,505
1980	1,595,138
1990	1,631,766
2000	1,953,631
2010	2,100,263

Source: US Census Bureau.

from taxes.[61] Of course, Lyndon Johnson's efforts were not limited to the oil industry. The decision of the federal government to build the $60,000,000 NASA complex in 1962 in Houston provides billions in economic benefit to the region today and provides a link to the aerospace and aviation industry that diversifies the Houston economy.[62] As table 8.1 illustrates, the efforts of the Houston business community and success of Texas with the federal government made Houston a global energy powerhouse going into the 1970s.

Of course, the oil industry in Houston capitalized on the growing importance of the automobile in American society but, again, the creation of the interstate highway system by the federal government greatly increased the miles traveled by the average American in their gas-guzzling car. Houston's leadership in the early to mid-1900s truly understood how to capitalize on political connections to maximize the economic benefits of the booming oil industry. By capitalizing on market forces and utilizing connections with the federal government, Houston moved from a mid-sized American city to becoming a major American city.

Houston's Move from Major American City to Global Energy Center

Houston's move from a major American city in the 1970s to the global energy center it is today is just as interesting as its rise from a mid-sized city in the early 1900s. Again, calculated public policy moves, this time at the municipal and state level, positioned Houston to be the global energy powerhouse it is today. Houston's primary appeal to the energy industry and business in general is an overwhelming support for free enterprise and against government intrusion in the marketplace. "Free enterprise is still the gospel" is a famous quote about Houston in *Fortune* magazine in 1980, and it still remains the mantra today.[63]

Houston's Free Enterprise Zoning Approach

The clearest example of Houston's support for free enterprise over government regulation lies in the fact that Houston is the only major city in America with a traditional Euclidian Zoning Code. The US Supreme Court in the 1920s authorized cities to separate land uses through a locally enacted zoning code. The Euclid, Ohio zoning code in question divides land within a jurisdiction into districts, or zones, with varying restrictions on uses that may be established and conducted in the different zones and standards (such as size and location of buildings, yard areas and intensity) such uses must meet. Zoning regulations "regulate and restrict the height, number of stories, and size of buildings and other structures, the percentage of a lot that may be occupied, the size of yards, courts, and other open spaces the density of population, and the location and use of buildings, structures and land for trade, industry, residences and other purposes...The local legislative body may divide the municipality into districts of such number, shape, and area as may be best and within such districts it may regulate and restrict the erection, construction, alteration, repair or use of buildings, structures or land."[64] Traditional Euclidian Zoning creates neighborhoods with housing subdivisions; shopping centers with exclusive parking areas; office/business parks with exclusive parking areas; civic institutions, such as churches, schools, and libraries, separated from other uses and served by exclusive parking areas; and roadways, connecting these separated land uses and designed exclusively for cars.[65]

Houston does not follow traditional Euclidian Zoning; however, Houston uses a land use master plan and does regulate land use just not at the level of every other major American city. Houston's land use master plan goes back to the early 1900s and the city also regularly supports local land use issues by assisting with land assembly and infrastructure. Houston also regulates land use without the adoption of a formal zoning code. Houston divides the city up into two zones—urban and suburban.[66] The urban zone within the Loop 610 corridor permits high-density developments with small lot sizes and narrow streets.[67] Houston regulates minimum lot sizes, parking requirements, setbacks, street widths, block sizes, building lines, and prevailing size requirements, all of which impact commercial, residential, and industrial developments.[68] Houston home-owner associations use deed restrictions to manage land use through private agreements.[69] Houston's 80 "super neighborhoods" permit residents, civic groups, and businesses to develop a land use plan and set neighborhood priorities.[70] Houston is not without land use restrictions, but its approach contains substantially fewer restrictions and expedites the land use regulation process for private development.

Houston Benefits from Business-Friendly State of Texas Policies

Houston clearly benefits from multiple business-friendly policies adopted by the state of Texas. Three of these business friendly policies center on labor, tax, and annexation issues. State labor policies influence the unionization rate that a state enjoys. Many corporations are attracted to states with fewer labor unions. Unions present a challenge to business when negotiating worker agreements, but also are effective at increasing the wage rates of employees. While additional wages help workers in those companies, rising wages also contribute to a higher cost of doing business in a region or state.

Texas Labor Policy Impacts the Energy Economy

Houston benefits from the fact that Texas is a Right to Work state. Right to Work laws empower workers to decide whether or not to join a union even if a majority of the workers vote to form a union. The unionization of manufacturing companies in the Industrial Midwest and Northern parts of the United States helped Southern states adopt Right to Work laws and attract global companies interested in avoiding a labor union. The Southern part of the United States, including Texas, grew much more quickly than the rest of the United States, and is a leader in global manufacturing.[71] Part of the reason for the success of Texas and other Southern states in attracting manufacturing is that the lower wages they pay workers make the location of factories in these regions much more cost competitive with expensive states in the Midwest, West, and Northeast. Southern states pay nearly $2 an hour lower than the Midwest, $4 less than the West, and $7 less than the Northeast.[72] Houston is attractive to manufacturers not only because of the connection to the energy industry, but also because Texas' Right to Work law keeps wage rates lower than other US competitors and leads to workforce unionization rates nearly one-third of the national average.

Texas Tax Policy Supports Houston Energy Economy

Houston also benefits from the state of Texas' lack of state and municipal income tax. The state of Texas and its municipalities do not operate with income taxes. Seven states—Alaska, Florida, Nevada, South Dakota, Texas, Washington, and Wyoming—do not have an income tax. Non-income tax states experience higher economic growth compared to states with a state income tax.[73] However, all states and local governments provide services and raise taxes to meet public demands for roads, public safety, trash, water, sewers, schools, and other vital public services. Unlike the majority of other state governments and many municipalities across the nation, Houston, Texas does not charge its residents or those that work within the city an income tax. Tax policy impacts economic development. Houston and the state of Texas still charge taxes, primarily sales and property taxes, and they offer the same array of government services that states with an income tax offer. The state of Texas not only does not have a personal income tax, but it does not have a corporate income tax either.[74] The state operates instead on a Margins Tax that operates much like a gross receipts tax used in many other states. The state does not have a property tax but does permit local governments, local school districts, and special taxing districts to impose a property tax.[75] Property taxes in Texas in 2011 were estimated to total at $42,000,000,000 with an annual growth rate of 6.4 percent from 1989 to 2008.[76] School districts in Texas gain 55 percent of the total property tax collected.[77] The state of Texas does use a 6.25 percent sales tax on tangible personal property and certain services, and permits cities, counties, and transit authorities to add to the local sales tax rate—but only to a maximum of 8.25 percent.[78] As with most other states, the state of Texas also exempts many manufacturing activities and other specific industries from the application of state taxes. Residents and businesses in Houston do not pay an income tax, but companies pay a franchise tax (usually 1 percent); pay a total property tax

$2.52922 per $100 of appraised value individually to the City of Houston, Harris County, Houston Independent School District, and the Houston Community College; pay a 2.7 percent tax for general entry rate for the first $9,000 of earnings for unemployment insurance; and pay an 8.25 percent sales tax individually to the state of Texas, City of Houston, and the Harris County Metropolitan Transit Authority.[79]

Texas residents have a low tax burden, but its businesses rank eighteenth in the nation for tax burden.[80] Thus, Houston uses a wide array of economic development tax incentives at its disposal to retain and attract companies. Houston's tax incentives include public finance mechanisms, tax credit, and tax abatements.

Houston's Tax Increment Reinvestment Zones (TIRZs) finance the cost of development by setting aside taxes attributable to new improvements, known as tax increments, to fund public improvements within the zone.[81] A TIRZ can only be used in an area that is substantially harming the city or county's economic, public health, safety, morals, welfare, or housing growth because of substandard slums, deteriorated or deteriorating structures, defective sidewalks or street layouts, faulty lot layouts, unsanitary or unsafe conditions, the deterioration of site improvements, tax, or special assessment delinquencies, defective titles, conditions that endanger life or property, and structures rarely used for commercial, industrial, or residential uses.[82] Houston can implement TIRZ plans, acquire property, develop agreements to implement TIRZ plans to regulate land use, construct public works infrastructure, and develop educational facilities.[83] TIRZs are governed by a Board of Directors, which prepares and implements a project and financing plan for the zone.[84] Project plans need a map illustrating the existing and planned use for the zone, proposed zoning changes, a list of non-project costs, and an outline of whether people will be displaced by the plan.[85] The financing plan needs a detailed list of project costs within the zone, planned public works, an economic feasibility study, planned bond debt and monetary obligation schedule, financing method description, total appraised value of taxable real property within the zone, the estimated captured appraised value of the zone, and the duration of the TIRZ.[86]

Many examples of successful Houston TIRZs exist. Houston's Uptown TIRZ was created in 1999 to address the challenges of success of the Uptown area.[87] Uptown is less than 10 miles from Downtown Houston and is primed for economic development. The neighborhood already has $3,590,000,000 in gross retail sales, 33 hotels, nearly 200,000 residents, and 28,000,000 square feet of office space.[88] However, Uptown cannot succeed without additional infrastructure investment; thus, the Uptown TIRZ has developed a plan to address traffic congestion and promote new development to enlarge the tax base.[89] The Uptown TIRZ plan includes $235,000,000 in new improvements that should spur over a billion dollars in new development and $341,000,000 in new taxes.[90] The Uptown TIRZ road construction improvements under way are possible only with the Houston TIRZ program.

Houston also aggressively uses a property tax abatement program to retain and attract jobs, but Houston will not abate ad valorem taxes unless another local government offers to do so.[91] Houston's tax abatement program is targeted at high-employment facilities in distressed neighborhoods where the location of at least 25 jobs over three years can redevelop the community.[92] A $1,000,000 real estate investment is typically required and the company seeking the abatement must illustrate that if they are not granted an abatement, they will leave Houston.[93] Projects located in state designated Enterprise Zones reduce the minimum investment level to $500,000 and require the creation of only five jobs, but 25 percent of new employees must live within the Enterprise Zone or be economically disadvantaged.[94] Eligible projects include buildings, machinery and equipment, site improvements, leasehold improvement increasing the property value, office and related personal property needed to operate the facility, as well as efforts to modernize the facility. Ineligible property includes land, inventory, supplies, tools, furnishings, vehicles, vessels, aircraft, housing, property to be rented or leased, electric generation, and natural gas facilities.[95] 90 percent tax abatements for ten years will be used for targeted industry clusters of biotechnology, information technology, aerospace, and energy, new or expanding businesses in struggling neighborhoods, and transit-oriented development within 1,500 feet of transportation corridors.[96] Houston uses its tax abatement program regularly and is trying to use a $3,000,000 tax abatement to lure a major Chevron development to Downtown Houston.[97]

Texas Pro-Growth Annexation Policy Benefits Houston

Annexation is another tool used by local governments to further economic development. Annexation is a legal process used as a tool by municipalities to bring generally unincorporated land into its boundaries and foster the development of the annexed site.[98] Annexation becomes an economic development tool when the process is used to ensure that the annexed territory gains access to municipal services, generally, water, sewers, and roads, to ensure that a planned or existing development can happen. Economic development cannot happen without the provision of basic public services provided by local governments. In many states, municipalities are the form of government that can provide the largest array of public services. Many municipalities will only provide these services if the unincorporated land is annexed or becomes a part of that municipality. The municipality will gain the tax revenue from the new territory and the new territory will receive the public services that only the municipality can provide. Annexation can be a win-win for municipalities and economic developers. However, those nearby residents opposed to development or local governments, such as townships that lose out on tax revenue when their territory is annexed into a city, are opposed to the use of annexation. As state governments draft the rules, local governments and developers must follow to implement a municipal annexation, many states choose to empower the residents and local governments opposed to annexation.

Texas state law is supportive of the use of municipal annexation and is part of the reason Houston is a global economic powerhouse. Texas municipal rights related to property exist within its boundaries and the extraterritorial jurisdiction area immediately surrounding the city. Houston's 5-mile band around their boundary constitutes its extraterritorial jurisdiction, and Texas law gives Houston the ability to annex land within its extraterritorial boundary. State law gives Houston the ability to annex land by general-purpose annexation and limited-purpose annexation.[99] General-purpose annexation requires Houston to meet public notification requirements, and all affected property become part of the general-purpose boundaries, giving these residents all the rights and public services related to the city's citizenship but subjecting them to all city regulations and taxes as well.[100] Restrictions for general-purpose annexation are limited, including a declaring the intention to annex the property in the city's annexation plan at least three years prior to the annexation.[101] Limited-purpose annexation needs to be part of a Strategic Partnership Agreement with a utility district, does not need to be part of a three-year annexation plan, and requires less public notice. [102] It applies primarily to commercial properties.[103] However, though limited-purpose annexation does not apply to property taxes, it does apply to municipal sales tax on retail sales.[104] SPA annexed territories do not require a three-year planning mandate, but do mandate which regulations and services will apply to the territory.[105]

Texas annexation policy is pro-growth and has helped Houston grow into an economic powerhouse; for instance, it permitted Houston to annex the 50,000 residents of Kingwood. Houston not only is 600 square miles, but the city has even reserved 1,300 square miles of land for future growth.[106] Texas' aggressive annexation policy has empowered Houston to have land available for development.

Houston Infrastructure Vital for Region's Energy Economy

Houston would not be a global energy center without its massive public infrastructure. Houston's infrastructure impacting their global economy starts with the Ports of Houston, but also includes the Houston Airport System and substantial rail and highway systems. The Ports of Houston includes facilities in Houston, Galveston, Freeport, and Texas City.[107] The Port of Houston, otherwise known as the Houston Ship Channel, is the top ranked American port in global tonnage and is the largest Gulf Coast container port.[108] This 52-mile-long channel handles two-thirds of all Gulf Coast container traffic and is well positioned for additional growth with the planned expansion of the Panama Canal, as larger ships from the Pacific will have easier access to Gulf Coast ports.[109] 150 companies utilize the Houston Ship Channel moving more than 200 tons of cargo annually, carried by 8,000 ships and 200,000 barge calls.[110] The Port of Houston is an economic juggernaut. The Houston Ship Chanel contributes 1,026,820 Texas jobs and generates over $178,500,000,000 in Texan economic impact.[111] The Houston Ship Channel generates more than $4,500,000,000 in state and local tax revenues.[112] As previously discussed, Houston is the nation's top exporting city and this would not be possible without the Houston Ship Channel.

The Houston Ship Channel generates all this economic impact because it transformed Houston from not just the place where oil and natural gas was discovered, but the place where these resources were refined and processed and distributed to the world. The development of a logistics play matched with the oil and natural gas resource and refining ability made Houston a global energy powerhouse. Innovation that developed larger and larger container ships to bring oil into the United States and move oil out of the United States were met by Houston with efforts to consistently invest in the Houston Ship Channel to ensure this resource remained a major asset.

Ships are not the only thing to travel through Houston and aid its economic development. The Houston Airport System serves over 50,000,000 travelers and is the eighth busiest international passenger American airport system.[113] The Houston Airport System serves 190 destinations—121 of these are nonstop domestic destinations and 69 direct and nonstop foreign destinations from George Bush Intercontinental Airport.[114] Houston is also served by William P. Hobby Airport that has directs flights to more than 40 domestic destinations.[115]

As with the Port of Houston, the Houston Airport System impacts economic development. The Houston Airport System's total economic impact in 2010 was measured at over $27,500,000,000; that included direct, indirect, and induced impacts.[116] Direct impacts include production of passenger, cargo, government and private air transportation services, and average annual capital expenditures.[117] Indirect impacts constitute the local spending by air visitors and induced impacts are the spending and re-spending by recipients of income due to the direct and indirect impacts.[118] The Houston Airport System employs 234,000 people generating $8,800,000,000 in direct, indirect, and induced earnings.[119] The global Houston Airport System provides a launching pad for regional business leaders to market to the world, as well as a landing site for global investors to arrive and do business in Houston. In particular, Houston's strong base of direct flights into Central America is matched by only a few regions.

Houston is also a global leader in rail traffic, with more than 800 miles of rail line and 21 miles of railroad bridges.[120] Ten major rail companies serve Houston region and work an impressive distribution system with 150 trucking lines connecting the Port of Houston to the United States, Canada, and Mexico. Houston's three Class I Railroads operate more than 96 percent of Texas' Class I tracks.[121] Houston also operates a world-class light rail network.

Finally, Houston's network of roads, highways, and bridges include a crossroads for Interstate Highways 10 and 45 and Interstate 610, US 59, US 90, US 290, Texas 99, Texas 146, Texas 225, Texas 249, Texas 288, Hardy Toll Road, Westpark Tollway, and Beltway 8.[122] The Houston region's 4,206.41 lane miles of freeway and expressways use over 100 miles of high occupancy vehicle lanes.[123]

Houston Workforce Key to Future Success

If Houston's infrastructure investments in the early twentieth century led to 100 years of growth, the regions workforce development investments may be the key to success in the twenty-first century. 10,000 Americans are retiring every day. The Baby Boom generation is leaving the workforce and the demands of a modern workplace—whether an office or factory—are higher than ever before. All regions of the United States are winning or losing the economic development battle, in large part based upon how well they replace their aging workforce. Houston's overall employment is at record highs—nearly 3,000,000 workers live in the Houston region. 36 percent of these workers have a college or associate degree.[124]

As figure 8.2 illustrates, management, business, science, and arts lead all occupations in Houston, constituting 35 percent of all workers in the region, followed by sales and office occupations at 23.5 percent, service workers at

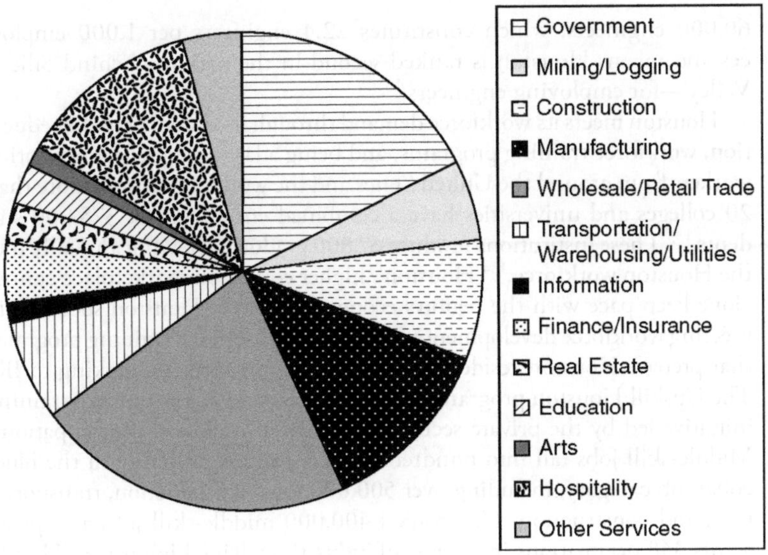

Figure 8.2 Houston employment by industry.

Source: Greater Houston Partnership, Workforce Fact Sheet, 2014, retrieved from http://www.houston.org/newgen/10_Employment_and_Workforce/10B%20W001%20Employment%20by%20Industry.pdf.

17 percent, natural resources, production, transportation, and materials-moving jobs at 12 percent, and construction and maintenance occupations at 11 percent.[125] Houston's employment in 2012 amounted to nearly 25 percent of all the jobs in Texas and an equal percentage of the state's management, business, and production jobs.[126] Considering the size and scale of the state, that job total on a statewide basis is substantial. The professional and business services sector in Houston is on the rise as demand for engineering and computer systems design increases.[127] Trade, transportation, and utilities' growth is more related to the general local consumer growth of the region, and the education and health services sectors, like all regions in the United States, benefits from regional population expansion and an aging population respectively.[128] 125,000 health care jobs alone have been added since 2012.[129]

Regions like Houston are focused not just on replacing their retiring workforce but also on developing high-wage jobs through a STEM workforce. STEM workers are 5 percent of the American workforce but equal more than 50 percent of the nation's economic growth.[130] STEM job growth is expected to exceed demand for non-STEM jobs.[131] Houston is a national leader for the location of STEM jobs, being home to nearly

60,000 engineers, which constitutes 22.4 engineers per 1,000 employees and means Houston is ranked second in the nation—behind Silicon Valley—for employing engineers.[132]

Houston meets its workforce demand through its system of higher education, workforce training programs, and being a large magnet for outstation workers from around the United States and the world. Houston's more than 20 colleges and universities have a combined enrollment of 375,000 students.[133] These institutions produce 67,800 graduates flowing annually into the Houston workforce.[134] Houston's higher education marketplace cannot alone keep pace with the region's economic growth. Houston succeeds in meeting workforce development needs through a series of training programs that prepare Houston residents for the jobs companies are looking to fill. The Upskill Houston program is one such example of a workforce training initiative led by the private sector to fill jobs in middle-skill occupations. Middle-skill jobs fall into hundreds of occupations primarily in the blue-collar job category, including over 500,000 jobs in production, transportation, and construction. Houston's 1,400,000 middle-skill jobs are spread across 348 occupations in a range of industries paying high wages. Upskill Houston is a business led six-point strategy to fill these middle-skill jobs. Strategies include formation of industry sector councils to gain key input on job needs; launching of a public awareness campaign on importance of middle-skills career opportunities; adopting common curricula to assess individuals and provide training in basic skills and employability; facilitating connections between stakeholders; developing tools for demand-side and supply-side data; and establishing educational and industry networks to disseminate sharing of best practices and industry partnerships.[135]

Houston also meets its substantial workforce needs because it is a magnet for global workers. As table 8.2 shows, Houston is an attractive market for Hispanics.

Table 8.2 Comparison of Houston's population by race

Race	Houston (in percent)	Texas (in percent)	US (in percent)
White	25.6	45.3	62.6
Black	23.7	11.8	13.2
American Indian	0.7	0.7	1.2
Asian	6	3.8	5.3
Pacific Islander	0.1	0.1	0.2
Two/More Races	3.3	2.7	2.4
Hispanic	43.8	37.6	17.1

Source: US Census Bureau Quick Facts.

Hispanics constitute the majority of Houston's population and nearly double the national average. Houston meets its workforce demand by being a global center to which Hispanic workers relocate.

Lessons Learned from the Houston Energy Economy

Houston is America's and maybe the world's global energy center operating a massive energy based economy. The question is how they did this and what lessons mid-sized cities that are yearning to become global energy centers can learn from them. Houston offers lessons for other cities trying to determine how to grow and succeed based upon an energy economy. Houston made the challenging jump from mid-sized city to mega-city. America has few mega-cities. Houston, Los Angeles, San Francisco, Seattle, Chicago, Atlanta, Chicago, Boston, and Washington, DC are America's current mega-cities, with populations in the millions. One hundred years ago, Houston was not on the list of American mega-cities. St. Louis, Boston, Baltimore, Pittsburgh, Cleveland, and Buffalo all made the top spots in the list of largest American cities in 1900.[136] Mid-sized cities becoming mega-cities do so based upon their proximity to industry, a growth in Latino population, increased educational attainment of their residents, and the annexation of territory.[137]

Houston ran by the 1,900 industrial cities by riding the energy wave. The discovery of oil at Spindletop was the earth's gift to Houston, but the growth of the oil industry tied to industrial production and the expansion of the automobile market positioned Houston for future growth. To Houston's credit, they did not rest on the economic benefits of oil production. Instead, Houston understood and recognized the need for manufacturing and distribution jobs to complement oil production and processing. Houston understood the link between manufacturing and distribution and secured strategic investments in their distribution infrastructure by building the Houston Ship Channel. The earth did not give Houston a port—Galveston had that. However, Houston built a port using impressive political influence led by a committed business community that now produces billions in economic returns and is the leading exporting site in the United States.

Houston also gained in the educational attainment of its population, in part, through the diversification of its workforce with the building of the NASA Johnson Space Center to launch a global aerospace center and the recruitment of the Texas Medical Center—again, by business leaders—to build the world's largest hospital and a major research and development center. Houston's ranking as the second largest population center for engineers, behind only Silicon Valley, illustrates the success of the region in attracting highly educated workers. Finally, Houston welcomed Hispanic workers and today shows a Hispanic majority population. All of Houston's positives—energy resources, strategic infrastructure investments, and STEM company and workforce recruitment—would still not be sufficient to make Houston the fourth largest city in the nation and a global energy powerhouse. Houston's economy would fail without the growing population of Hispanic workers that actually keep the city running.

Finally, few cities in America enjoy growth through annexation like Houston. Houston capitalized on a growth friendly Texas annexation law to the fullest extent possible. Houston's geographic size is simply enormous. Matched with its growth friendly approach to zoning, Houston aggressively uses land use planning strategies to develop the region.

Urban regions connected to an energy resource have an opportunity to become a mega-city over the next several decades. However, that opportunity is far from a guarantee. Houston offers compelling lessons and a guide for how regions, through smart business and government leadership, can capitalize on that energy resource and build a global energy economy.

Big Questions

1. What is the benefit of becoming a global energy center?
2. How did the market for automobiles impact Houston's path to become a global energy center?
3. How did politics impact Houston's path to become a global energy center?
4. How did land use policy impact Houston's path to become a global energy center?
5. How can another mid-sized city near an energy resource become the Houston of the twenty-first century?

Class Writing Assignment

Remember, GEC is pushing Nevada to address its workforce issue. Las Vegas cannot become an energy jobs center without additional college education and trained workers. GEC proposes to create an energy workforce strategy in partnership with the energy industry. Addressing the region's lack of college-educated graduates is a major challenge to the state's effort to recruit advanced manufacturing, services, and technology jobs in or out of the energy industry. Representative Phil Smith introduced the Nevada Energy Workforce bill to create an industry-based training program with $50,000,000 appropriated. Much debate is expected. The Tea Party and UNLV are advocating against the use of taxpayer dollars for this purpose while energy companies and unions are supportive. Representative Smith, Chairman of the House Education and Workforce Committee, will draft the legislation, present sponsor testimony, hear from proponents and opponents, and the committee will then vote on the measure.

Class Debate

Remember, GEC supports the development of a Detroit Energy Center with the goal of creating a self-sufficient energy district and other incentives to serve as a prime location for energy-related manufacturers and service companies. The Detroit Energy Center is proposed in partnership with the University of Michigan and the Detroit Energy Company on 2,000 acres of formerly tax delinquent land to build a world-class energy based research and development facility. The site will be energy self-sufficient and could be the spark Detroit needs to succeed. City Council must approve of the zoning for this project and neighbors to the project—Tea Party advocates and the region's natural gas company object. Supporters include the Mayor, Council President, and Detroit Energy Company support the project. The Detroit City Council President plans to introduce a resolution creating the Detroit Energy Center as one of Michigan's Renaissance Zones and the Detroit Energy Company's plan for development of renewable energy sources to make the site energy self-sufficient. Council will hear sponsor testimony from the President, as well as testimony for proponents and opponents, and then vote.

Additional Readings

Bartels, Christopher E. and Jeremy L. Hall, "Exploring Management Practice Variation in Tax Increment Financing Districts: Toward Administrative Theory of Performance," *Economic Development Quarterly*, 26, 1 (2012): 13–33.

Craig, Steven, *Houston Airport System Economic Impact Study, Final Report* (Jenkintown, PA: GRA, Incorporated, June 30, 2011).

Eichenthal, David R., *Why Mid-Size Cities Grow and What It Means for Chattanooga* (Washington, DC: Brookings Institution Center on Metropolitan Policy, September 2005).

Houston, Scott, *Municipal Annexation in Texas, "Is It Really That Complicated"* (Austin, TX: Texas Municipal League, March 2012).

Jarmon, Christopher A., James M. Vaderleeuw, Michael S. Pennington, and Thomas E. Sowers, "The Role of Economic Development Corporation in Local Economic Development: Evidence from Texas Cities," *Economic Development Quarterly*, 26, 2 (2012): 124–137.

Lacker, Jeffrey M., *Manufacturing in the New Southern Economy* (Roanoke, VA: Federal Reserve Bank of Richmond, Southern Growth's 2011 Chairman's Conference, June 13, 2011).

Oklahoma Council of Public Affairs Eliminating the State Income Tax in Oklahoma: An Economic Assessment (Oklahoma Council of Public Affairs, Arthur Laffer, November 2011).

An Overview of Taxes in Texas, Texas Taxpayers and Research Association, January 2011, retrieved from http://www.houstontx.gov/ecodev/tirz.html.

Shah, Neil and Ben Casselman, "'Right-to-Work' Economics: States That Bar Mandatory Union Dues Tend Toward More Jobs but Lower Wages," *Wall Street Journal*, December 14, 2012.

Vojnovic, Igor, "Governance in Houston: Growth Theories and Urban Pressures," *Journal of Urban Affairs*, 25, 5 (2003): 589–624, 603.

Chapter 9

Building a Green Economy

Chapter Goals

1. Understand the economic benefits of renewable energy.
2. Recognize the role of federal incentives in building an energy led economic development strategy.
3. Recognize the role of state energy mandates and incentives in building an energy led economic development strategy.
4. Be prepared to discuss how Oregon is an energy led economic development leader.
5. Be ready to discuss how Oregon became an energy led economic development leader.

The shale boom based upon the fossil fuels of oil and natural gas is not the only source of energy upon which an energy led economic development initiative is being built. Regions and states across the United States are building successful energy led economic development initiatives based upon renewable energy sources. Wind, biomass, solar, hydro, geothermal, and other renewable sources of energy are widely available across the United States; and regions with access to these energy sources are adopting policies, regulations, and launching retention and attraction campaigns, all geared toward increasing the use of renewable energy sources, and recruiting companies connected to the renewable energy sources. Renewable energy sources can supply 482,247,000,000 kWh of electricity annually, and this constitutes 118 times the nation's electric consumption rate.[1]

Before measuring the success of renewable energy led economic development, the definition of a renewable energy job needs to be established.

The United Nations have defined renewable energy jobs as serving a wide range of industries including agriculture, manufacturing, construction, installation, maintenance, scientific, technical, administrative, and service related occupations that contribute to preserving environmental quality.[2] These jobs protect and restore ecosystems and biodiversity and reduce energy, water, and material consumption by making efforts to reduce the carbon pollution resulting from the consumption of fossil fuels.[3]

As previously mentioned, renewable energy is gained primarily from wind, biomass, solar, hydro, and geothermal projects. The reason so many regions and states focus on renewable energy led economic development is that it serves the environment, national security, and economic development purposes.

Green Benefits of Renewable Energy

The use of renewable energy is better for the environment. Taking energy from the wind, water, or the sun does little to pollute the earth. Taking energy from fossil fuels leaves pollutants as a by-product of the energy consumption process. Many regions focus on renewable energy primarily to address larger air, water, and land environmental pollution concerns. High-population regions such as California have been forced to look at renewable energy as one of the means to address the challenging air quality the millions of cars produce for their region.

One-third of all US carbon emissions are a result of electricity production from fossil fuels sources such as coal and natural gas.[4] Natural gas emits between 0.6 and 2 pounds of carbon dioxide per kWh and coal emits between 1.4 and 3.6 pounds of carbon dioxide per kWh.[5] On the other hand, wind emits only 0.02–0.04, geothermal 0.1–0.2, and hydroelectric between 0.1 and 0.5 pounds of carbon dioxide per kWh.[6] Mandating that renewable energy sources make up 25 percent of America's energy sources by 2025 would lower carbon dioxide by 277,000,000 metric tons annually.[7] Replacing fossil fuels with renewable sources would have public health benefits, as the costs of fossil fuel pollution on Americans health care costs have been estimated to be in the billions of dollars.[8]

Energy Security and Renewable Sources

Energy security is another reason renewable energy sources are popular. Fuel diversification and use of more domestic sources of energy have long been a part of America's national energy policy. The Arab Oil embargo in the 1970s enhanced the profile of this approach. Renewable energy sources are popular because they involve the use of domestic sources of energy that are available in large, if not infinite, supplies. A region's wind, water, and sun offer energy sources that last as long as the earth itself, as it is replenished on a regular basis. Energy diversification is also popular, as it prevents consumers and the economy from being attacked by shocks in the energy market.

Again, the Arab oil embargo illustrated the economic challenges of having nearly 100 percent of American's transportation industry dependent upon one energy source—in this case, oil. The automobile boom fueled the need for foreign oil, which created political and economic leverage among nations in the Middle East against the United States. Middle-Eastern oil was turned into an economic and political weapon against America for their political and military support of Israel. A lack of diversification of types of energy and the nations from which America gains the energy illustrates how energy security becomes an issue.

Renewable Energy Jobs

Renewable energy is also popular because of its economic development benefits. The renewable energy industry constitutes nearly 3,000,000 American workers in high-wage jobs.[9] A larger percentage of jobs in the renewable energy industry are in the sought-after manufacturing industry, and these same companies export twice as much as the typical American company.[10] Renewable energy manufacturing includes companies producing electric vehicles, green chemical products, lighting segments, and biofuels.[11] Renewable energy jobs not only pay wages 13 percent higher than the average US jobs, but they are available for low and middle-skilled workers left out by the migration of many American manufacturing jobs.[12]

The renewable energy industry is in growth mode as well. US demand for renewable energy equipment is estimated at $14,000,000,000–$20,000,000,000 through 2030.[13] This is roughly the size of America's construction machinery, farm machinery, and household appliance manufacturing industries.[14] 70 percent of the jobs from wind energy are expected to be in the manufacture of wind energy components.[15] Wind energy manufacturers such as General Electric, Acciona, Clipper, Gamesa, Nordex, Siemens, and Vestas, and their Tier I suppliers are major targets for economic growth for regions investing in wind energy.[16] It is hoped that 50 percent of American wind projects will use domestically built wind energy components.[17] Ease of logistics and transportation costs are seen as the main bases for economic optimism for American windmill production. It is difficult to transport large wind generators over long distances by ship or rail.

Wind energy is the most promising of Michigan's renewable energy sources. Wind energy in the United States grew by 25–50 percent annually in a range of states, including Texas, Iowa, California, and many others.[18] Wind is of course popular because of its environmental benefit of not producing any negative pollution, but it concerns rural residents due to noise issues and its negative impact on bird populations. Also, compared to traditional fossil fuels, wind energy is not as consistent a source of energy, as it is dependent upon minimum wind speeds.

Wind energy creates direct, indirect, and induced economic benefit. Direct economic benefits are generally small. A typical $200,000,000 wind energy project will create 100 MW of power and provide 6–8 permanent jobs to operate and maintain the windmills.[19] However, these windmills provide property tax payments to local and state governments ranging from $3,000 to $12,000 per MW landowner leases of $3,000–$7,000 per MW.[20] However, indirect economic benefit can be substantial, as regions with many windmill locations are likely to be the locations for windmill manufacturing and production, as the windmill generators are so large that long-range transportation is difficult. The extended or induced economic benefit of the windmill industry is likely to be over 100,000 jobs as this manufacturing industry begins to grow in the United States.[21] The total economic impact of two large wind farms in Texas was $1,800,000,000.[22]

Oregon offers a national model for the successful development of a renewable energy led economic development model. Oregon's renewable energy led economic development model is based upon a strategic plan, utilizes regulations and incentives to promotes the use of renewable energy and works to retain and attract companies in the renewable energy industry. Oregon's foray into renewable energy began in the 1970s and continues strong today. Oregon gets a large percentage of its power from renewable sources, with hydro projects at the top of the power production list. This state has long promoted the use of renewable energy sources as the state's primary means to gain energy. Oregon's renewable energy strategy is working. Oregon's 43,000 renewable energy jobs with nearly 5,000 employers constitute 3 percent of the state's workforce.[23] Covering all major industry and occupational labor groups, Oregon's renewable energy jobs pay $23.07 an hour, which is nearly $7 an hour more than the state's median wage.[24]

Build It and They Will Come

No matter the state or region, renewable energy led economic development strategies are built on far more than tax incentives structured to be attractive to companies in this industry. Renewable energy led economic development is far more complex and involves controversial regulatory and incentive decisions that could drive up the cost of power for all consumers in an attempt recruit renewable energy companies. The focus of renewable energy led economic development is on the production and consumption of energy from both existing and new energy sources, and then uses incentives and regulation to promote energy diversification and the creation of reliable, low-cost energy alternatives.[25]

First, renewable energy led economic development strategies adopt a series of regulations designed to promote the use of renewable energy sources. Mandates are often placed on utilities to require their use of a certain percentage of their energy to come from renewable sources. These mandates are controversial, but are designed to serve as a statement to the renewable energy industry that their state is open for business with this industry. These renewable energy use mandates are as much about preserving and protecting the environment as they are about economic development. Mandates can also be placed on development growth patterns to adopt "smart growth" plans to reduce energy consumption and promote mass transit services. Finally, mandates are placed on the construction and building industry to require the adoption of "green building" strategies.

Going along with the "stick" is, of course, the carrot when it comes to renewable energy led economic development. A long list of local, state, and federal incentives exist to provide tax credits, tax abatements, grants, and loans as rewards for installing and using renewable energy devices. Also, incentives are crafted and used to retain and attract companies in the renewable energy industry through tax credits, tax abatements, and workforce and brownfield grants. These mandates and incentives are seen as an effort to promote the building of renewable energy facilities to encourage renewable energy manufacturers, research and development centers, and company headquarters to come to that state.

Federal Government Support for Renewable Energy

The federal government is a major supporter of renewable energy. Federal tax credits can be taken for eligible renewable energy systems including:

- Solar—30 percent tax credit for solar energy expenditures excluding passive solar systems and solar pool-heating systems;

- Fuel Cells—30 percent credit for fuel cell expenditures capped at $1,500 per 0.5 kW of capacity;
- Small Wind Turbines—30 percent tax credit for small wind turbine expenditures up to 100 kW in capacity with a maximum credit of $4,000;
- Geothermal Systems—10 percent tax credit for geothermal expenditures, including heat pumps and equipment used to produce, distribute, or use energy derived from a geothermal deposit;
- Microturbines—10 percent tax credit for microturbines at $200 per kW of capacity for microturbines up to 2 MW in capacity with an electricity-only generation efficiency of 26 percent or higher; and
- Combined Heat and Power (CHP)—10 percent tax credit for CHP property systems up to 50 MW in capacity that exceed 60 percent energy efficiency, excluding CHP systems using biomass for at least 90 percent of the system's energy source.[26]

Beyond tax credits, the federal government is a major provider of funding to support the development of renewable energy led economic development. Seven federal agencies support renewable energy with loans, market rate debt, below market rate debt, tax credits, grants, payments, credit enhancement, and performance contracting. Federal agencies that support renewable energy include the federal Department of Energy, Housing and Urban Development, Treasury, Environmental Protection Agency, Transportation, Agriculture, and the Small Business Administration.[27]

Federal loans to support renewable energy are a major source of funding for companies and organizations looking to promote energy led economic development. The federal Department of Energy's Loan Program provides loan guarantees to renewable energy projects and direct loans to advanced technology manufacturers of advanced technology vehicles and components.[28] This program offers more than $30,000,000,000 in loans, loan guarantees, and commitments, supporting more than 30 projects.[29] Project examples includes large wind and solar farms, thermal energy storage systems, nuclear reactors, and auto manufacturing facilities. Tesla Motors, Inc. offers an example of the Department of Energy's Loan Program success.[30] A $465,000,000 federal loan helped Tesla build their widely successful all electric plug-in vehicle. 1,500 jobs are tied to this loan and its success in the marketplace permitted Tesla to pay back their entire loan nine years early.[31]

Federal Energy Department loan guarantees support in a limited number of core technology areas including advanced grid integration and storage, drop-in-biofuels, waste to energy, energy efficiency, and efficiency improvements.[32] Advanced grid integration focuses on renewable energy systems such as demand response or local storage to demonstrate grid generation compatibility of generation from renewable resources.[33] Drop-in-biofuels centers on making current engines, delivery infrastructure, and refueling station equipment more biofuel friendly. Waste to energy funding looks to landfills or ranches as a source for methane.[34] Enhancement of existing facilities projects are funded to incorporate renewable generation technology into existing renewable energy and efficient energy facilities.[35] Examples of existing facility projects include incorporating power production into non-powered dams, the use of variable speed pump-turbines into hydro facilities, and retrofitting wind turbines.[36] Finally, efficiency improvement projects are geared toward promoting the use of innovative technology to reduce greenhouse gas emissions in residential, institutional, and commercial facilities and buildings, and in energy storage and processing.[37] Advanced fossil energy projects are another focus of the Department of Energy loan program. Again, there are technology areas of focus that include advanced resource development, carbon capture, low carbon power systems, and efficiency programs.[38] The focus of the advanced fossil energy project program is to improve existing systems in order to improve their environmental performance.[39]

The US Department of Agriculture also supports the development of renewable energy projects. The Electric Loan Program funds the operation of the nation's rural electric network. The federal government backs the operation of 600 electric systems in 46 states serving 40,000,000 customers.[40] The Rural Utilities Service's Energy Efficiency and Conservation Loan Program helps finance a range of energy efficiency programs, including grid renewable energy systems, demand side management projects, energy audits, consumer outreach, power factor correction equipment, and re-lamping to more energy-efficient lighting.[41] The Rural Utility Service also funds the High Energy Cost Grant program to purchase energy generation, transmission, or distribution facilities in communities with energy costs at least 275 percent higher than the national average.[42] This program funds renewable energy projects as well as energy efficiency and energy conservation projects.[43]

The Treasury Department $3,200,000,000 in Qualified Energy Conservation Bond program permits local and state government and tribal communities to fund energy conservation projects.[44] The goal of the program is to reduce energy consumption in public buildings by at least 20 percent.[45] This is low cost, financing option is out of funding but was popular when available.

The SBA CDC/504 Loan Program provides financing for major fixed projects that can be used for energy efficiency projects. Numerous examples of SBA CDC/504 Loans being used for energy efficiency exist.[46]

Colorado's Rocky Mountain Excavating, Inc. utilized an SBACDC/504 Loan to support an energy-efficient facility's expansion in Castle Rock.[47] This disabled-veteran-owned small business and certified 8(a) small business provides general contracting, construction management, and design-build services throughout the Rocky Mountain region.[48] The Department of Agriculture's Business and Industry Loan Guarantee Program finances rural companies through a loan guarantee for business and industrial acquisitions to keep the business from closing; encouraging job expansion, purchase of land, building, or equipment; or up to $25,000,000.[49] This program can be used for the purchase of energy-efficient equipment and may be combined with other programs.[50]

The federal Department of Agriculture Water and Waste Disposal Direct Loan and Grant program develops water and waste disposal systems in rural areas smaller than 10,000 people.[51] Communities must be unable to obtain financing from other sources and federal funding can be used for construction, land acquisition, professional service fees, interest, equipment, initial operation and maintenance costs, and project contingencies.[52] Similarly, the federal Water and Waste Disposal Guaranteed Loan Program provides loan guarantees for the construction or improvement of water and waste disposal projects serving the same rural communities with less than 10,000 people.[53] Guaranteed loans are made by lenders to public not-for-profit corporations and Native American tribes.[54]

Another Department of Agriculture program is the Biorefinery Assistance Program that supports new and emerging biofuel technologies.[55] This program provides loan guarantees as high as $250,000,000 to develop commercial-scale biorefineries and biobased product manufacturing facilities.[56] Projects must convert renewable chemicals with biomass and biosolids.[57] The Advanced Biofuel Payment Program provides payments to producers to support production of advanced biofuels refined from sources other than corn.[58] Eligible renewable biomass crops includes cellulose, sugar and starch (other than corn), hemicelluloses, lignin, waste materials, biogas, butanol, diesel-equivalent fuel, sugarcane, and nonfood crops such as poplar trees or switch grass.[59] The Rural Energy for America Program supports agricultural producers and rural small businesses with loan guarantees and grants to fund renewable energy systems and energy efficiency improvements.[60] The Energy Audit and Renewable Energy Development Assistance Grant Program provides grants for energy audits to agricultural producers, defined as companies or organizations producing crops, livestock, forestry products, hydroponics, nursery, and aquaculture products, where at least 50 percent of their income is derived from the agriculture product.[61]

The USDA also operates the Repowering Assistance Program that pays biorefineries to use renewable biomass as a replacement for fossil fuels.[62]

Project funding will be awarded based upon the availability of funds, project scope, and the amount of biomass to be used.

Infrastructure is another major area of focus for federal renewable energy funding.[63] The Transportation Infrastructure Finance and Innovation Act of 1998 created a federal credit program for eligible major surface transportation projects through secured (direct) loans, loan guarantees, and standby lines of credit.[64] The program's goal is to support innovative funding partnerships with state departments of transportation, transit operators, special authorities, local governments, and private entities.[65] This program provides a funding opportunity for transit projects that cost a minimum of $50,000,000, where federal funding does not exceed 33 percent of the project total.[66] The list of federal renewable energy programs is long and the application process can be complex. However, many successful regional and state renewable energy strategies are built in part with federal funding as their basis.

State Renewable Energy Strategies

State governments provide a wide range of regulations and incentives to support the use of renewable energy and promote the location of renewable energy companies in their states. Energy source disclosure is a common tool used to support renewable energy. Some states require electrical utilities to provide their customers with specific information about the electricity that the utility supplies.[67] This information, which generally must be shared with customers periodically, usually includes the utility's fuel mix percentages and emission statistics.[68] In states with restructured electricity markets, generation disclosure policies are designed to help consumers make informed decisions about the electricity and suppliers they choose.[69] Some states permit government, businesses, residents, schools, and nonprofit organizations to buy electricity from renewable resources or by buying renewable energy credits.[70] Many state and local governments, as well as the federal government, have committed to buying "green power" to account for a certain percentage of their electricity consumption.[71] Green power purchases are typically executed through contracts with green power marketers or project developers, through utility green power programs, or through community aggregation.[72]

Some states require utilities to provide information regarding renewable energy options to customers, and some even require utilities to offer their customers the option of paying more for green power.[73]

Many states created public benefit funds during the electrical utility restructuring era in the 1990s, in order to develop a pipeline of funds for renewable energy programs.[74] These public benefit funds take a small surcharge from every customer and designate it for renewable energy programs. The renewable energy strategy attracting the most attention is the state's adoption of Renewable Portfolio Standards (RPSs), which mandates that utilities use or buy renewable energy or renewable energy credits for a certain percentage of their electricity sales.[75] Often, RPS establishes percentage mandates for various renewable energy sources such as solar, wind, biomass, and hydro.[76]

To promote customer use of solar and wind facilities, some states enact policies giving residents and businesses the right to install and operate a solar or wind energy system and preventing local governments from regulating or banning these facilities.[77] In addition, state and local policies may be adopted to give "sun" rights to residents and businesses looking to use solar facilities.[78] Also, some states have established a licensing process for solar and wind energy contractors to ensure high-quality installation and maintenance of these systems.[79] Many of these same communities may adopt local construction-permitting rules to move solar and wind and other renewable facility projects to the "top of the stack" in order to expedite the construction of these facilities.[80] Some states have capped fees that local governments may charge for a permit for a solar or wind energy system.[81] Finally, many states adopt a wide range of tax incentives—including abatements, credits, grants, and loans—to encourage the installation of renewable energy facilities as well as to retain and attract companies in the renewable energy industry.[82]

Oregon's Renewable Energy Led Economy

Oregon is a national model for how to build a renewable energy led economy. This state benefited from population migration from California and grew by transforming their economy, once dominated by timber and agriculture, into a renewable energy powerhouse. Nothing about Oregon's weather or geography makes renewable energy easy. The region is famous for its rainy weather, lacks a major mega-city, and is wedged between the major tech leaders in Seattle and San Francisco. Rather than mimic those regions, Oregon instead went the route of renewable energy and their approach produced economic results. Their approach was not an accident or the result of one company hitting it big. Instead, Oregon took the 1970s energy crisis to heart and started a focus on renewable energy at that time. Oregon uses both incentives and regulations to promote the renewable energy industry and to encourage the use of renewable energy sources and devices to support the industry they are recruiting. Oregon also benefited from what the earth gave them and invested in renewable hydropower along its many rivers flowing from the mountain ranges in the eastern part of this beautiful state. Oregon proves public policymakers can design and implement a strategy that promotes renewable energy and develops jobs.

The Oregon Economy

Oregon's population is approaching 4,000,000, which makes it a mid-sized state, but it is a state with a growing population.[83] Oregon's general economic story is mixed. Its per capita income is just below the national average but average home value is just above the national average.[84] Oregon's 29.2 percent college graduate rate puts it just above the national average with an average travel time to work that is lower than the national average.[85]

Oregon's general economic data is good but not great. However, Portland, as the largest city in the state and its economic leader, has stronger economic numbers than the rest of Oregon. Portland grew 14 percent since 2000, and is a young city with a median age of 36.[86] Portland's population is also better educated than the rest of Oregon, and far better than most of the rest of the United States. 44 percent of Portland is college educated, which puts it far above the national average that hovers near 30 percent.[87] Portland's median income is higher than the rest of Oregon at around $31,000 and above the national average, and this city's unemployment rate is a full point lower than the rest of Oregon.[88]

Oregon's Renewable Energy Led Economy

The renewable energy industry is the impetus behind Oregon's economic revival. Oregon is a fast-growing regional American renewable energy economy. Oregon gained more than 19,000 jobs in the renewable energy sector between 1998 and 2007.[89] The Oregon renewable energy economy comprises a larger share of the Oregon economy than any other state in the union, and these jobs grew nearly seven times faster than other jobs between 1998 and 2007.[90] The nearly 2000 Oregon renewable energy companies attracted nearly 100,000,000 in venture capital and filed nearly 200 patents just from 2006 to 2008.[91] Renewable energy companies such as Abundant Renewable Energy, a leader in small scale wind turbine systems construction, and Vulcan Power Company, a leader in constructing and manages geothermal energy facilities, are just a few of Oregon's renewable energy victories.[92]

Oregon consistently ranks high for nearly every renewable energy industry award. Oregon ranked second to California in a recent State Clean Energy Leadership Index and beat out Massachusetts, New York, and Colorado, who rounded out the top 5 states.[93] Oregon's high rank among renewable energy officials is based upon a number of factors. 64 percent of the state's energy comes from in-state electricity generation that comes from utility-scale renewables, biomass, and hydro projects.[94] Oregon ranks #3 among the states for hybrid electric vehicle adoption on a per-capita basis and the state ranks first in the number of EV charging stations per capita.[95] Oregon also ranked #1 among the states for LEED-certified green building projects per capita and finished second for smart meter penetration—representing 25 percent of total meters.[96]

Oregon's Renewable Energy Led Economic Development Strategy

Oregon utilizes a "carrot and stick" approach to promote a renewable energy led economic development strategy. Oregon's renewable energy led economic development strategy addresses utility regulatory policies, creates financial incentives for renewable energy and energy efficiency programs, and develops land through a "Smart Growth" approach. Oregon uses both the "carrot" and the "stick," from a policy standpoint, all with the goal of retaining and attracting renewable energy companies.

Oregon's Regulatory Renewable Energy Led Economic Development Strategy

Oregon is aggressive when it comes to its regulatory approach with utilities and renewable energy and energy efficiency. From a regulatory or "stick" approach, large Oregon utilities are required to use 25 percent of their energy from renewable sources by 2025, small utilities have a 10 percent goal, and the smallest utilities have a 5 percent goal.[97] Specific solar power mandates are also established. Oregon utilities have a solar sourcing minimum requirement of 20 MW by 2020, but solar systems operating prior to 2016 are credited 2 kWh for each generated kWh.[98]

Oregon also permit large utilities to use unbundled renewable energy credits sourced locally for 20 percent of their 25 percent by 2025 renewable energy sourcing mandate to meet up to 50 percent of their obligations.[99] In addition, the use of municipal solid waste as a renewable power source is permitted, but is restricted to 11 MW of power each year.[100] Utilities are also permitted to pay an alternative compliance payment if they cannot meet their renewable energy mandate.[101]

Net metering for most Oregon utilities is capped at 2 MW for non-residential private utility customers and 25 kW for municipal utilities, co-op, and public utility district customers.[102] A net metering total cap of 0.5 percent of utility's historic single-hour peak load for municipal utilities, co-ops, and public utility district consumers also exists.[103] Oregon also established interconnection standards for most investor-owned utilities that set system capacity limits for greater than 20 MW for large generators, less than 10 MW for small generators, greater than 25 kW for residential net metering, and 2 MW for non-residential net metering.[104]

Oregon's Renewable Energy Incentives Offer the Carrot to Green Industry

Oregon also operates with a wide range of renewable energy incentives to promote the use of renewable energy as well as to recruit renewable energy companies. Oregon's Biomass Producer or Collector Tax Credit supports the use of biomass as an energy source.[105] The biomass tax credit goes to agricultural producers or collectors for biomass used for biofuel, anaerobic digesters, pellets, or torrefaction to produce energy.[106] The biomass may be in a solid, liquid, or gaseous state, but the biomass must be produced and used in Oregon.[107] Firewood, charcoal, construction and demolition debris, urban wood, yard debris, sawdust, algae, and material from pre-construction or construction activities and golf courses are not defined as biomass for purposes of this program.[108] This transferable tax credit varies per biomass type and rates are multiplied by the amount of biomass produced or collected and vary by biomass type as follows:

- Oil seed crops at $0.05/pound;
- Grain crops at $0.90/bushel;
- Virgin oil or alcohol at $0.10/gallon;
- Used cooking oil or waste grease at $0.10/gallon;
- Wastewater biosolids at $10/wet ton;
- Woody biomass at $10/dry ton;

- Vegetative biomass from crops at $10/dry ton; and
- Animal manure or rendering offal at $5/wet ton.[109]

Oregon's Tax Credit for Renewable Equipment Manufacturers provides a 50 percent credit equal to the construction costs for a renewable energy manufacturing facility up to $40,000,000.[110] The credit applies to companies that manufacture systems that harness energy from wood waste or other wastes from farm and forestlands, non-petroleum plant or animal based biomass, the sun, wind, water, or geothermal resources.[111] Oregon's Energy Conservation Tax Credit provides up to a 35 percent tax credit for renewable thermal energy project costs over five years.[112] The project's first year of energy savings must yield a simple payback period of greater than three years.[113]:

Oregon local governments can set up Rural Renewable Energy Development Zones.[114] Only regions outside the urban growth boundary of a large city are eligible.[115] Companies meeting job creation and capital investment requirements are eligible for a 3–5-year local property tax exemption.[116] Eligible technologies include wind, geothermal, solar, biomass, or biofuels.[117] Zones will have locally set tax abatement caps with a $250,000,000 abatement ceiling.[118]

Oregon is focused on the retention and attraction of renewable energy equipment manufacturers and offers a targeted tax credit to support these companies. Companies engaged in the manufacture of solar, geothermal heat pumps, wind, biomass, small hydroelectric, tidal, or wave energy equipment are all eligible for the Oregon Renewable Energy Equipment Tax Credit.[119] A 50 percent tax credit rolled out in 10 percent increments over five years can be awarded to these companies up to a total incentive of $20,000,000.[120] It covers facility construction costs as well as machinery and equipment purchases for renewable energy manufacturing facilities for existing or new facilities.[121]

Oregon offers the BETC to commercial, industrial, residential, nonprofit, schools, and local and state government agricultural and institutional groups.[122] This performance-based incentive program provides a tax credit equal to $0.0856/kWh for 10 years for photovoltaic systems greater than 10 kW in capacity up to 200 kW.[123] Some solar power customers benefit from the Solar Volumetric Incentive and Payments Program that provides payments to solar customers for permanently installed solar systems over their useful life.[124] Payments vary based upon the size of the system and the participating utility.[125]

Oregon's tax credit for energy conservation projects for commercial buildings involves renewable thermal, commercial, agricultural, and industrial processes, "Cool Schools," and combined heat and power.[126]

The energy payback period cannot be longer than three years and the five-year tax credit equals 10 percent of the project costs in years one and two and 5 percent for the balance of the credit.[127] Costs covered by this tax credit include professional service firm fees, title searches, escrow fees, permits, licenses, construction materials, and labor by contractors.[128] The more energy saved the larger the tax credit that will be awarded.

Oregon offers property tax incentives for solar, photovoltaics, landfill gas, wind, hydro, fuel cell, geothermal, and methane gas projects for commercial, industrial, and residential customers.[129] The Oregon property tax incentive shields from taxes the real market property value gain achieved based upon the installment of renewable energy equipment.[130] The property tax exemption applies to end-user customers with net-metered projects intended to offset on-site electricity use.[131] Oregon offers commercial and government customer's renewable energy facility grants as well. These small $50,000 grants are targeted at electricity-generation projects of 25 kW–10 MW, and are funded from a legal settlement with Reliant Energy.[132] Oregon provides grants for feasibility studies for renewable energy projects in preparation for participants in Oregon's Small-Scale Energy Loan Program.[133] This program is a revolving loan fund designed to be paid back.

Oregon's electric-utility restructuring law mandated the state's main utilities create a public-purpose charge of 3 percent from customers' bills to support renewable energy projects.[134] The Energy Trust of Oregon was designated to administer a range of programs including:

- Oregon Business Energy Efficiency Rebate for Existing Buildings program is offered from commercial, government, multifamily, certain high-energy consumption retail, Data Centers, and other projects to provide energy efficiency incentives capped at $500,000 per site;
- Industrial and Agricultural Production Efficiency Program provides funding for technical assistance and cash awards to cover the costs of industrial process energy efficiency improvements to address lighting, green motor rewinds, heat pumps, variable speed drives, new HVAC equipment, irrigation systems, and compressed air improvements;
- New building energy efficiency incentives for new commercial building construction or major renovation projects;
- Grant Writing and project development assistance for 20 MW or less renewable energy projects to support grant writing, feasibility studies, or technical assistance with design, permitting, or utility interconnection paying up to half the cost associated with project development activities to a maximum of $40,000 and up to $150,000

in project incentives to cover the delta between project costs and the energy's market value;[135]

- Oregon's Solar Electric Incentive Program supports customers of Pacific Power and PGE who install new photovoltaic systems on new or existing residential, commercial, and government facilities; and
- Oregon's Small Wind Incentive program supports Portland General Electric and Pacific Power customers' installations of up to 50 kW turbines with incentives ranging from $5.00 per kWh for systems with annual energy output of 9,500 kWh or less or $47,500 plus $1.75/kWh for systems with an annual energy output over 9,500 kWh.[136]

Most Public Benefit Fund projects must be located in the service territory of Pacific Power, Portland General Electric, NW Natural, and Cascade Natural Gas to be eligible to participate.[137]

Oregon created a Renewable Energy Development Grant Program for commercial and residential customers, which provides up to $250,000 or 35 percent of the total project costs from tax credit auctions.[138] A relatively small total budget of $3,000,000 exists for this program.[139] Oregon's State Energy Loan Program is a 35-year program created by state constitutional amendment directing the state to financially support through the sale of bonds to finance small-scale, local energy projects.[140] This program offers low-interest loans for energy saving and producing projects from renewable sources for individuals and businesses alike, and loans range from $20,000 to $20,000,000.[141]

Oregon's Sustainable Land Use Strategy

Oregon is the home of "Smart Growth." As a reaction to the migration of California residents, Oregon is an early adopter to managing its population growth through land use regulation. Oregon's approach moved from the top down and was led by the state government. State government leaders chose not to let the multiple local government leaders create a development path. The result is a state with an integrated planning system.[142] Oregon's state and local governments work in tandem to develop a comprehensive land use plan and this plan needs to be consistent with statewide planning goals.[143] The Oregon Land Conservation and Development Commission develops the state's land use master plan,[144] and is empowered to enforce this land use plan by withholding grant money or revenues from communities out of compliance with the master plan.[145]

Oregon Metropolitan Service Districts also play a large role in land use master planning efforts.[146] As an example, the Portland Metro not only provides regional services, but it also works on land use, transportation, parks and open spaces, water management, natural hazards, and implementing measures for an urban growth boundary.[147] Portland Metro manages growth in the multi-county region and creates a sustainable development strategy appealing to renewable energy companies.[148] Oregon's Smart Growth land use strategy promotes density in development and is appealing to the energy efficiency and renewable energy company base.

How Does Oregon Afford Their Renewable Energy Led Economic Development Program

Oregon built an energy led economic development economy the "old fashioned way"—they bought it with regulations and incentives. The real question for others trying to learn from Oregon is how the state can afford this strategy. High energy costs drive away manufacturers almost as fast as organized labor. A 3 percent tax on utility customers, even if it is used for energy efficiency programs, creates added costs for residential and business consumers alike. How is Oregon successfully attracting the manufacturers to the state with these additional energy costs?

First, Oregon understands renewable energy companies are attracted to regions and states that promote their industry through the use of renewable energy and energy efficiency equipment and services. General Motors would never locate an automobile manufacturing factory in a state that did not support the use of cars. The same holds true for the renewable energy industry.

Second, Oregon benefits substantially from low-cost and green hydropower. Oregon uses renewable sources of energy for a substantial amount of its power needs, but this is not coming from windmills or solar panels. The fast running rivers covering Oregon and substantial capital investments in past years create a currently cheap form of power.

Smart regulatory decisions also help Oregon promote renewable energy. The utility regulatory structure in Oregon created cheap power and millions of dollars in renewable energy incentives. Also, Oregon benefited from a population shift from California to Oregon and other states on the California border. Oregon does not have to fight for population growth. Instead, Oregon is managing population growth and managing it in a way the migrants from Northern California are completely used to.

Finally, Oregon started a renewable energy led economic development strategy when Jimmy Carter was President and they have stuck with it.

A long-term focus by Oregon policymakers is paying off for twenty-first century Oregon.

Big Questions?

1. What elements are needed for a successful renewable energy led economic development initiative?
2. Can states and regions keep traditional manufacturing and attract renewable energy manufacturing companies?
3. What pitfalls need to be avoided when building a renewable energy led economic development initiative?
4. What made Oregon a successful renewable energy led economic development region?
5. Can Oregon be replicated by other regions and states?

Class Writing Assignment

Wind Michigan needs to build their case that wind power is a critical piece of the state's energy infrastructure. They have asked you to write a 500-word op-ed article for the Detroit Free Press newspaper defining the economic and environmental benefits of wind energy for the state of Michigan. Wind Michigan expects the op-ed piece to define what wind energy is; how it compares with other sources of an energy from a technical, availability, and resource standpoint with other energy sources; how much power it produces in Michigan; what the potential future growth for wind in Michigan is; and what is the potential economic impact, based upon studies from similar projects proposed in Ohio.

Class Debate

Wind Michigan is opposed by the incumbent investor-owned electric utilities as being too expensive. These electric and natural gas utilities are promoting a reversal of the Michigan Renewable Energy Portfolio Standard. They claim Michigan's mandate that the state's energy come from a percentage of renewable sources will drive up costs substantially and they promise a 10 percent rate cut if the standards are removed. The Michigan

Chamber of Commerce supports the utilities, but the renewable energy and environmental community is opposed. State Representative Tom Edison proposed the utilities legislation, and the Michigan House Public Utilities Committee will review the legislation, hear Representative Edison's sponsor testimony, proponents, and opponents, and vote on the bill.

Additional Readings

American Council on Renewable Energy, Oregon Fact Sheet, 2014.

Clean Edge Clean Energy Oregon Report, 2012.

Guide to Federal Financing for Energy Efficiency and Clean Energy Deployment (Washington, DC: Department of Energy, September 2014).

Lantz, Eric, Frank Oteri, Suzanne Tegen, and Elizabeth Doris, "State Clean Energy Policies Analysis: State Policy and the Pursuit of Renewable Energy Manufacturing," *National Renewable Energy Laboratory Technical Report NREL/TP 6A2–46672*, February 2010.

Meck, Stuart (ed.), *American Planning Association, Growing Smart Legislation Guidebook: Model Statutes for Planning and the Management of Change, Note 6C: A Note on Existing Regional Plans* (Chicago, IL: American Planning Association, 2002).

Muro, Mark, Jonathan Rothwell, and Devashree Saha, *Sizing the Clean Economy: A National and Regional Green Jobs Assessment* (Washington, DC: Brookings Institution, July 13, 2011).

Pew Center, Oregon Clean Energy Economy Report, 2008.

Slattery, Michael C., Becky L. Johnson, Jeffrey A. Swofford, and Martin J. Pasqualletti, "The Predominance of Economic Development in the Support of Large-Scale Wind Farms in the U.S. Great Plains," *Renewable and Sustainable Energy Review*, 16 (2012): 3690–3701.

Slattery, Michael C., Eric Lantz, and Becky L. Johnson, "State and Local Economic Impacts from Wind Energy Projects: Texas Case Study," *Energy Policy*, 39 (2011): 7930–7940.

Chapter 10

Shale Energy Revolution

Chapter Goals

1. Understand how shale oil and natural gas impacts the American economy.
2. Recognize the connection between developing rural regions and shale energy.
3. Understand the "boom and bust" elements of shale energy.
4. Be prepared to discuss the short-term strategies to prevent shale energy ghost towns.
5. Be ready to discuss the long-term strategies to prevent shale energy ghost towns.

Thurmond, West Virginia was a bustling, small town in South-Central West Virginia, in the middle of Coal Country. Like many small towns, Thurmond was built on the backs of the railroad that lived and died on the excavation and distribution of coal. Founded in the mid-1800s, Thurmond was incorporated as a city in the early 1900s and, at its peak, was home to nearly 500 residents in 1930.[1] The town had two hotels and a thriving core of small businesses.[2] Thurmond is located on the scenic New River and a bridge crossing the river gave Thurmond its strategic location. The passenger rail service was operated by Chesapeake and Ohio Railway (C&O Railroad), and the region was part of a global coal-mining operation.[3] However, a fire that destroyed one of the two hotels in town in 1930 started the decline, which never stopped. Today, Thurmond has five residents left, the train depot is a national park service center, no post office, and the jobs and commerce are gone.[4] Thurmond is one of hundreds of ghost towns left by the rise and fall of the energy industry.

Thurmond Is a Story of the Challenge of Rural America

Rural America struggles for economic survival. 17 percent of the US population lives in rural communities, with the Western United States constituting the largest land mass of low-density rural land.[5] While rural communities enjoy a high quality of life, low crime, and good schools, they have been struggling to survive economically in the twenty-first century economy. Farming, once the dominate employer in rural communities produces more crops than ever, but the technology that changed American manufacturing also impacted agriculture's role as a large employer. In 1970, agriculture constituted 13 percent of high-density rural employment and 23 percent of the jobs in low-density rural communities.[6] By 2007, those employment numbers for agriculture dropped to 6 percent and 12 percent respectively.[7] The service, government, retail, and manufacturing sectors are larger employers in rural communities.

This reduction in employment is an outgrowth of the tremendous efficiency of the American agriculture industry. Farm productivity nearly tripled while all other industrial productivity increased by 75 percent from 1900 to 2000.[8] In fact, in 1900, 41 percent of the total American workforce was employed on farms—by 2000, that number was down to just 2 percent.[9] Productivity of the American agriculture industry is good news for consumers and owners of farms; however, this productivity left many rural centers searching for economic success.

Rural communities on average provide lower incomes, higher poverty, less educated populations, and higher mortality rates than their urban counterparts. Per capita income in urban communities increased by 38.9 percent from 1979 to 1999, but only increased by 35.2 percent in high-density rural communities and 29.9 percent in low-density rural communities.[10] Rural communities are also struggling from a demographic standpoint. On average, rural communities have a higher, older population. The young people are growing up and leaving. Rural communities also lack broadband access that a younger generation simply presumes will exist. 60 percent of rural Americans do not use broadband services (10 percent less than urban households), often because their telecommunications service falls short of their urban counterparts.[11] Rural communities lack the young workforce most companies are seeking. Poor rural communities also lack the infrastructure for development. Without growth and development, rural communities often lack traditional land use master plans, experience with public finance vehicles used for infrastructure prepared for massive growth. The shale boom town of Williston, North Dakota faces an existing $625,000,000 infrastructure bill just to keep up with the current pace of energy developments.[12]

Big thinkers were so concerned about the fate of rural America that they suggested turning the Great Plains back into natural fields where buffalo could be free to roam and raised as cattle.[13] This notion of a Buffalo Commons was promoted as a restoration project; like rural America is a contaminated "Brownfield," and needed to be restored back to its natural state to find a productive use.

These struggling rural communities are often the source of great energy resources, and the lack of population density makes mining that resource possible in ways an urban center could not imagine. Downtowns of major cities with large land improvements are not attractive sites for an oil-drilling rig. That does not of course mean states like Oklahoma hesitated at all to place an oil-drilling rig right next to its State Capitol. However, for the most part, urban centers cost too much and cause too many headaches for energy companies to place large-scale wind farms, horizontal drilling rigs needed for shale exploration, coalmines, or traditional vertical oil rigs. Thus, energy companies are attracted to rural communities where property is relatively cheap and ready for drilling from day one.

Opportunities of the Energy Boom

The energy industry's discovery of energy in a rural community creates a tremendous economic boom. Look at North Dakota as an example of the boom. The explosive growth associated with the Bakken Shale region in parts of North Dakota are changing this rural, farming state's economic future. North Dakota is now the second largest oil producing state in the nation—up from ninth—the lowest unemployment rate, fastest growing economy in the nation, and a $2,000,000,000 state budget surplus.[14] North Dakota is exploding with millionaires as well as oil. This Midwestern state moved up 14 spots in an annual ranking of states by number of millionaires, moving from forty-third in 2012 to twenty-ninth, just ahead of the retiree-rich state of Florida.[15]

Williston, North Dakota is a small community at the center of North Dakota's shale boom.[16] For three years running, Williston is the fastest growing micropolitan community in the nation, having tripled in size through annexations, enjoyed a multi-million dollar community college expansion, and boasting a new city recreation center.[17] Apartments and commercial and retail centers are sprouting up like weeks. Recently a $500,000,000 Williston retail center was announced.[18] Williston, North Dakota and many other small rural towns are on the upswing of the shale oil boom.

North Dakota is not alone with the current shale gas energy boom. The Eagle Ford Shale region in a fourteen-county region of Texas is producing billions of dollars of economic benefit. This South Texas region gained $19,200,000,000 in economic benefit—as measured in 2011—that created 38,000 full time jobs supporting nearly $10,500,000 in gross regional product, $211,000,000 in local government tax revenue, and $312,000,000 in state tax revenue.[19] It is estimated that, by 2021, the Eagle Ford Shale could produce over $60,000,000 in economic output, support over 80,000 jobs, and generate over $3,000,000,000 in local and state taxes combined.[20] Communities such as Karnes County, Texas have seen $3,277,139,310 in economic output with an opportunity to triple that by 2021.[21] Employment growth in this fourteen-county region grew by 9.2 percent from 2009 to 2011—this during the worst economic recession in decades.[22] Half of that job growth was in the natural resources and mining industry occupation.[23] Local and state government also benefitted from a boom in the sales tax. In 2002, the Eagle Ford shale region provided just under $600,000 in sales tax, but, by 2012, spurred triple digit sales tax growth locally.[24]

The same boom is seen in Pennsylvania. The Marcellus Shale developments are a billion dollar bonanza for the Keystone State. 2011 studies indicate that the Marcellus Shale provided over 23,000 Pennsylvania jobs and just over a $3,000,000 economic impact that included over a $1,000,000 in worker income.[25] Pennsylvania enacted a local impact fee law that permitted local governments to collect a development fee related to the explosion of shale. Overall, the Pennsylvania impact fee collected $225,000,000 in 2012 and over $600,000,000 since 2012.[26] Many rural counties located in the Marcellus Shale boom gained millions in impact fee revenue, including just over $7,000,000 for rural Bradford County, Pennsylvania.[27] Bradford County, Pennsylvania drilled more shale wells than any other county in this shale rich state. Statewide, oil and gas companies paid $2,000,000,000 in corporate income tax to the state of Pennsylvania.[28]

The small towns of North Dakota, Texas, Pennsylvania, and elsewhere are leading a global energy revolution. The economic boom for these small communities and the states that house them is unprecedented in these communities.

The Energy Boom Brings a Bust

However, the reality of the energy industry is with a boom comes a bust. The energy industry is dynamic with steep rises and falls often caused by technological development, over-supply, and policy shifts in Washington, DC. The story of the coal industry offers a telling example relevant for today's boom towns. Coal for a variety of scientific, market, and regulatory reasons was king of the American energy market. Coal is a highly energy-rich resource that, when burned, produces the heat needed to produce energy much better than other sources. Coal is also plentiful in the United States with several regions of the nation being coal rich. Thus, when the industrial revolution hit full steam, it was fueled by oil and coal. Fast forward to the 1970s. Precipitated by an Arab-Israeli conflict, the Arab oil-producing nations in the Middle East launched an oil embargo on Israel's biggest supporter—the United States. America's post-World War II economy had made the nation and its consumers rich, and owning one or more cars in a household was central to the American dream. The federal government, in response to this auto boom, built a national inter-state highway system that altered American development patterns to move the population more out than up. Thus, suburbs sprawled all over the place.

The Middle East oil embargo hurt the American economy and turned the 1970s into the age of the energy policy. A major reaction to the oil embargo was substantial regulatory changes made to discourage the use of oil and promote the use of coal as an alternative fuel for heating and cooling homes, commercial, retail, and office centers. Thus, coal became king again. Coal company earnings increased steadily by about 150 percent from 1969 to 1977, and average earnings for workers increased by 5.5 percent.[29] Unfortunately, as the oil embargo faded and the commodity price of oil dropped, the coal market dropped as well. Coal jobs simply evaporated from 1982 to 1989, and earnings per day worked actually dropped and, in some regions, the number of coalmining jobs decreased from 37.4 percent to 17.6 percent.[30] Coal jobs have suffered because technology made this industry more productive while requiring fewer workers. Coal productivity continues to rise, but the overall employment in the coalmining industry falls as a result. In 1940, 130,000 West Virginians worked as coalminers, but that number is now down below 21,000.[31] The near future is not good for coal either. By 2015, West Virginia's coal production is expected to drop by 40 percent.[32]

The coal producing states of West Virginia, Ohio, Kentucky, and Pennsylvania all had regions suffer economic hardship due to the decline of the coal industry. West Virginia has not recovered from the decline of coal. The state has the fourth lowest per capita income, above-average unemployment rate, and the seventh highest poverty rate in the United States, and is actually losing population.[33] West Virginia's decline is linked to many factors, but the main one is the state's reliance on the success of the coal industry. In 2012, jobs in the mining industry saw a net decline in West Virginia, in large part due to the decline of coal jobs. [34] All this is bad news for the coal industry workers, but also for those dependent on the West Virginia state and local governments. The coal industry just a couple years ago provided 10 percent of all local and state government revenue even as the industry continued to decline.[35]

Surviving the Energy Boom and the Bust: Think Short and Long Term

The boom and bust of the coal industry is relevant to all regions winning the current energy revolution. The Williston's North Dakota, Karnes County, Texas, Bradford County, Pennsylvania, and other shale boom communities may see shale gas production dwindle from the ground. Over production can drive prices down for natural gas and oil, which halts not just the exploration for additional shale gas wells but also stops the flow of oil and natural gas from existing wells. To prove this point, production in the boom town of Bradford County, Pennsylvania slowed due to the drop in price for natural gas.[36] An accident could shift public opinion against shale gas exploration, much as the Three Mile Island incident stopped the American nuclear energy industry in its tracks.

Small towns in rural communities are rarely ready for massive energy-based developments and struggle to keep up with the demand for infrastructure even during the boom times. Williston, North Dakota has over $600,000,000 in local infrastructure needs just to keep up with current activity.[37] That is the bad news. The good news is that the development of energy resources produces new and unprecedented levels of wealth in a community and presents these previously struggling regions with a big opportunity. It is an opportunity that is not likely to last forever. Energy resources can dwindle. Once-popular sources of energy have now become unpopular for public policy reasons or they because too costly due to competitive sources, new technology, or regulatory changes. However, the boom town does not have to become a ghost town when the region's energy source goes bust. Proactive strategies launched during the boom times can prevent total economic collapse during the bad. Proactive strategies launched during the good times include both short-term and long-term economic development action steps.

These action steps are all shaped by basic or building block strategies involving land use, workforce, and infrastructure issues. The more advanced Five Drivers of Economic Development strategies geared toward recruiting high-wage jobs in growing industries centered on economic diversification strategies to recruit advanced manufacturing and technology companies are also needed to diversify these energy-based economies, in order to prevent a bust from becoming an economic collapse.

Short-Term Economic Development Strategies around an Energy Boom

Surviving a boom time may be harder than it sounds. Rural communities with no recent history of economic growth are rarely prepared with essential governmental tools to manage and plan growth, fund infrastructure, and prepare for massive real estate transactions. Communities enjoying expansion from shale developments illustrate the short-term challenges from an energy boom. From Ohio to Texas to North Dakota to Pennsylvania, rural small town America at the center of the shale drilling is working hard to manage the growth and development. Without a large pool of qualified workers at the ready, most of these communities have been invaded by out of state workers who have few options for housing. While hotels and apartment buildings are popping up all over shale country, these residential developments simply are not keeping up with the demand for housing. Instead, oil and natural gas companies desperate to capture and distribute oil and natural gas build "man-camps." These temporary housing options are not meant to last and are not built for the traditional comforts of the American home. Man-camps became a service offered by third party vendors to develop mobile, corporate housing. It sounds better than it is but the reality is the oil and natural gas industry could not run without man-camps. Small, rural communities with a new energy asset such as shale are not prepared for the influx of thousands of workers. Short-term strategies to survive an energy boom in a small, rural community address land use and economic development planning and infrastructure funding. Failure to address these issues results in either communities over growth with nothing but short-term, low quality developments that will not survive the test of time, or a complete stoppage of the energy boom because the drilling and processing facilities cannot gain access to the roads, water, sewer, and solid waste facilities that keep an energy project running. Availability of water is a critical aspect of the shale drilling process.

Plan for Current and Future Growth

Housing is just one piece of infrastructure that must be addressed in the short term for small, rural communities to capitalize on the growth and development of a new energy resource. While difficult to do, rural communities at the center of the energy revolution need to adopt an economic development master plan that maps out a land use road map and identifies key industries for which the region will work to retain and recruit. Local government manages design, growth, and development through the adoption of a land use or comprehensive plan. This legally binding document sets the guiding policies for city council's land use decisions. A comprehensive plan also guides private sector developers looking to make investments. A comprehensive plan is based upon research and knowledge of the community, and uses text and maps to identify problems and opportunities. Comprehensive plans regulate land use to meet the overall economic development goals of a community. If the community wants to promote its Downtown as a center for commerce, the comprehensive plan outlines how the Downtown will be zoned for permitted businesses, parking, and infrastructure to meet that goal. Finally, comprehensive plans use a system of evaluation to determine its overall success.[38]

The adoption of comprehensive plans are a democratic process following five general principles, including being future oriented; a continuous process; based on present and projected conditions; a fair and open community process; a comprehensive plan: not just something that covers individual issues such as sanitary conditions but takes a broader view of the future of the city.[39] The adoption of a comprehensive plan for a community is an essential step to providing a road map for the public and the private sector to understand how the community plans to succeed in the boom and bust times. Rural communities hoping for the location major industrial and commercial facilities based upon the availability of low-cost power need to adopt land use master plans that recognize that the size, scale, and scope of the industrial zoned sites are large enough to serve these facilities, and they are located with the roads, rail, water, sewer, power, telecommunications, and other infrastructure required for these facilities.

Land use planning does not guarantee the retention and recruitment of high-wage energy jobs. Pro-active economic development strategies are required as well. Small, rural towns searching for high-wage jobs, attracted to the energy industry and low-cost power, need to go beyond the traditional land use master plan and conduct an economic development industry cluster analysis to determine what growing industries are likely to locate in their region. Industry cluster strategies begin by identifying a region's business strengths through an analysis of the geographic concentration of firms within a particular industry.[40] This includes core firms and contributing organizations (lawyers, accountants, public relations, university faculty and others) the core firms need to survive.[41] Census and business data from reporting services or direct surveys—and economic data of the community—are the basic research required for an industry cluster analysis.[42]

Industry cluster strategies mandate not only the identification of the local industry cluster, but also a commitment to removing obstacles harming cluster growth, clearly defining the cluster boundaries, involvement of industry cluster participants, and private sector participation—all geared toward the creation of the cluster as an institution in the rural community.[43] Following the identification of the industry cluster, a supply chain analysis is conducted to expand the cluster participants. A market analysis then provides the growth opportunities for the industry cluster followed by the development of a Strengths, Weakness, Opportunities, and Threats (SWOT) analysis to connect a region's growth opportunities through a competitive lens. The SWOT analysis forces the rural community to take a hard look at not only their existing industries, but also to look at competing communities across the globe, in order to determine what industry cluster truly will succeed. Finally, a team is formed to implement plans to retain and grow the industry cluster identified.[44]

Address the Infrastructure Challenge

The number one immediate challenge facing shale boom towns is the infrastructure development not keeping pace with the rate of development. This is not a place where infrastructure is built into an industrial park waiting for development. True shale boom towns look like the Wild West during the gold rush. However, the boom slows if the community and companies cannot provide the roads, water, sewer, power, telecommunication, and rail service energy companies demand to capture and process this new energy resource. Energy companies themselves take the unusual step of building roads, turning lanes, installing traffic lights, and building other infrastructure needed for the massive truck traffic created by shale booms.

Local impact fees are a prime tool used to provide funding for the shale boom infrastructure. Pennsylvania offers an impact fee approach to helping local communities fund the infrastructure and other services needed by the shale boom. The Pennsylvania Oil and Gas Act of 2012 creates new impact fees collected from oil and gas operators, and the fees provided $202,000,000 in second year and $204,000,000 in the first year of the impact fees operation.[45] Annually, the fee may be adjusted upward based upon the Consumer Price Index if there is an increase in drilling activity over the previous calendar year.[46] Wells not producing natural gas in quantities greater than those of a stripper well (90,000 cubic feet per day) within three years have their fee suspended, but the fee is reinstated when the well produces gas in quantities greater than a stripper well.[47] Local governments are required to vote to impose an impact fee within 60 days of the act's effective date.[48] Impact fees are distributed to a wide range of state organizations prior to distribution to local governments and other state agencies, including conservation districts, Fish and Boat Commission, Pennsylvania Utility Commission, Pennsylvania Department of Environmental Protection, Pennsylvania Emergency Management Agency, Pennsylvania Office of State Fire Commissioner, Pennsylvania Rail Freight Assistance, and the Pennsylvania Natural Gas Energy Development Program.[49] Following allocations to these state agencies, the Pennsylvania impact fee awards 60 percent of its funding to local government agencies to fund affordable housing. The Pennsylvania impact fee provides direct awards to counties and municipalities impacted by shale development providing 36 percent to qualifying counties with unconventional gas wells, 37 percent to qualifying municipalities with unconventional gas wells, and 27 percent to qualifying municipalities in a county with unconventional gas well.[50] Municipalities cannot gain more than $500,000 or 50 percent of the total budget of the municipality for the prior fiscal year beginning in 2010 adjusted for inflation.[51] 40 percent of the impact fee revenue available after funding the initial state agencies and municipalities are awarded to various state agencies including 20 percent for the Commonwealth Financing Authority, 10 percent for the Environmental Stewardship Fund, 25 percent for the Highway Bridge Improvement Restricted Account, 12.5 percent for the Pennsylvania Infrastructure Investment Authority, and 12.5 percent for the Commonwealth Financing Authority—H2O PA Program.[52] Counties can gain 15 percent of this fee for planning, acquisition, development, rehabilitation and repair of greenways, trails, open space, natural areas, community conservation and beautification projects, community and heritage parks, and water resources management and funds are distributed to counties proportionally on a population basis.[53]

Public finance tools also provide local government with a method to fund infrastructure developments. Traditional public finance tools available to small, rural communities enjoying a shale boom include Industrial Revenue Bonds (IRB), Special Assessments, Transportation Improvement Districts (TID), and Tax Increment Financing. IRBs finance facilities and equipment for private001Esector organizations that qualify on a tax-free basis. Special assessments are a voted tax on property benefited by an improvement. TIFs are very popular with local government officials to finance public improvements where a local government pays for public improvements and infrastructure by capturing the future tax increments from the project's area.[54] The local government issues bonds to finance the project, and the bonds are paid for later through the increase in taxable property value that the improved area receives.[55] TIDs raise revenue for the repair of roads, highways, and bridges within a defined geographic area through the implementation of special taxes, tolls, and other fees.[56]

State governments, many times with the backing of federal government funding, offer a major source of infrastructure funding for booming, rural communities. Ohio offers multiple funding opportunities for local governments searching for infrastructure funding. As table 10.1 illustrates, sources of the funding include the Ohio Water Development Authority (OWDA), Ohio Development Services Agency (ODSA), Ohio Environmental Protection Agency (OEPA), and Ohio Water Authority (OWA), all offering programs that rural communities can use toward infrastructure funding.

During 2013, the OWDA funded infrastructure projects in 60 of Ohio's 88 counties, devoting over $600,000,000 to 244 local community infrastructure projects.[57] The OWDA Fresh Water Fund awarded $143,639,636 to 84 projects, the OWDA Community Assistance Fund awarded just over $100,000,000 to 43 projects, and the OWDA Water Pollution Control Fund awarded nearly $300,000,000 to 26 projects all over Ohio in 2013.[58] Of greater importance to the energy community, the OWDA awarded just over $4,000,000 in loans to Carrollton, Ohio in the heart of shale country.[59] This funding will assist in the development of a new water treatment plant for this 3,200-person city. Without a steady supply of water, shale development simply comes to a halt. For Carrollton

Table 10.1 Ohio infrastructure funding programs

Infrastructure Need	Program Name	Program Description
Drinking Water, Sewage, Water Pollution, Storm Water	OWDA Fresh Water Fund	Development of potable water sources, construction and expansion of water treatment, and distribution facilities
Drinking Water	OWDA Water Supply Revolving Loan Account Fund	Finance infrastructure to comply with Safe Drinking Water Act requirements
Sewage, Water Pollution, Storm Water	OWDA Water Pollution Control Loan Fund	Revolving construction loan fund for municipal sewage treatment plants
Water and Waste Water	OWDA Community Assistance Fund	\$3,000,000 loans for small community water or wastewater projects
Water and Waste Water System Design	OEPA Village Capital Improvement Fund	\$25,000 for planning and up to \$50,000 for design no-interest loans to low income villages for water supply and wastewater treatment projects
Water and Waste Water Projects	OWDA Rural Development Program	Interim financing at a short-term interest rates
Water and Waste Water Projects	ODSA Local Economic Development Program	\$5,000,000 loans related to a job-creation project
Sewage, Solid Waste, Water, and Hazardous Waste Projects	OWDA Private Industrial Bonds Program	Tax exempt financing for private companies engaged in public infrastructure development
Disaster Recovery	OWDA Security Assistance for Emergency Response Program	Low-interest local government loans up to \$100,000 for small communities and \$200,000 for larger communities to protect water and wastewater systems
Disaster Recovery	OWDA Emergency Relief Program	Relief from OWDA loans for local governments, where the locale is a federal or state declared disaster area and has incurred loss or damage to wastewater or water treatment facilities

Source: Ohio Water Development Authority, Program Summary, retrieved from http://www.owda.org/owda0001.asp?PgID=pi-summary.

to remain a boom town, it needs a reliable flow of water. The low-interest OWDA loan made this happen.

Development of Interstate Pipelines

A reality of developing shale gas is that, following extraction of the resource, natural gas and oil companies build facilities to process and distribute the gas across the nation. Processing facilities are often billion-dollar investments that produce jobs. Distribution pipelines are substantial investments as well, but they enable the companies to take the locally created resource and send it across the nation. These processing and distribution facilities make the oil and gas far more valuable, as the commodity is able to serve a national market and not just a local one.

The federal Natural Gas Act empowers the Federal Energy Regulatory Commission (FERC) to approve the construction and operation of natural gas pipelines.[60] While FERC lacks jurisdiction over the safety and security of pipelines, this federal agency certifies that the pipeline complies with US Department of Transportation safety standards.[61] The growth in FERC natural gas pipeline applications has occurred since the birth of shale gas. In the past, FERC may have had one or even zero natural gas line applications, but, in 2014, FERC has 13 of these requests pending and has already approved the installation of 7 pipelines.[62] FERC expedited the certification process by creating a pre-filing procedure to ensure that stakeholders were identified and engaged and concerns were addressed before the application was filed.[63] FERCs approval is based upon identifying the best route, with local support, for the natural gas pipeline. Building a pipeline involves traversing over the properties of many public and private landowners across hundreds of miles. While the natural gas company has eminent domain powers, the FERC is where the views of impacted property owners count.[64]

A critical legal power for utilities to build interstate pipelines is eminent domain authority. Eminent domain is the Constitutional authority to take private property for a public purpose while providing just compensation to the private property owners. Courts decide whether there is a valid public purpose and what the level of compensation should be. Eminent domain is guaranteed by the Fifth Amendment to the US Constitution, and state constitutions as well. State laws not only give eminent domain authority to local and state governments, but also permit utility companies to exercise eminent domain. Eminent domain guarantees private property rights in the United States, but also permits the government to take land for public facilities, roads, housing, economic development, and utility providers to construct and operate their networks.

Take the Long View for Energy Developments

Once the short-term issues of planning and infrastructure are addressed, small rural communities must address the longer-term economic development strategies to truly create the high-wage jobs that this new energy source creates an opportunity for. Long-term strategies include diversifying the regional economy beyond just the energy industry, capitalizing on new government revenue to enhance the regional quality of life, build stronger connections in the community through technology, build energy demand programs to create a new market for this energy source, and work with state regulators to ensure that the region and state are a low-energy cost market.

Moving Beyond Energy

To truly succeed, energy boom towns must diversify their economy beyond energy. This is just as true for Houston as it is for Williston, North Dakota. Diversification efforts need to be based upon an industry cluster analysis that identifies existing industries with strong growth prospects in the region. In addition, company retention and expansion campaigns for rural energy boom towns should center on industries dependent upon cheap power and industries connected to the energy resource. States developing a new, large pool of power and/or using regulatory or market tools to drive down the price of power have a unique opportunity to retain and attract high-wage jobs through reduced electric and natural gas power rates. However, just because a region has unique access to shale gas or wind power does not mean the region will gain jobs beyond the workers mining and processing this energy. States wishing to capitalize on new energy sources need to ensure that cheap power is available.

Step number one to building an energy led economy is promoting public policies supportive of the power industry extracting the resource. Areas where wind production can gain power from windmills should adopt policies promoting the siting and operation of windmills as well as regulatory policies friendly to the industry. Second, regions and states from a regulatory standpoint need to aggressively reduce the cost of power. Third, these regions need workers trained and prepared for energy industry and energy-intensive industry jobs. Industry based training programs provide pre-certified workers, ready to work from day one with an energy company or energy-intensive user.

Regions producing and offering cheap energy need to focus on energy-intensive industries. Company recruitment, workforce development, and site preparation should focus on the heavy industries of chemical, iron and steel, nonmetallic minerals, paper and pulp, and energy refining. These are the most likely long-term jobs to be created by the boom in local energy production. Energy-intensive industrial firms are following the American energy boom. Since 2011, over 100 new energy-hungry industrial plants have been announced in the Gulf Coast region alone, with $114,000,000,000 mostly centered in the petrochemical industry.[65] The American energy boom is even creating "back-shoring" opportunities. Methanex announced it was moving a methanol-producing unit back from Chile to Louisiana and a second plant is slated for startup in 2016.[66] Yara, a Norwegian fertilizer manufacturer, plans to jointly open a major ammonia plant with BASF on the Gulf Coast.[67] 19 new aluminum, iron, and steel plants have been announced following the availability of cheap energy.[68]

However, heavy industry will not just appear when asked to. States adopting business-friendly labor laws and tax friendly manufacturing policy will have more luck recruiting heavy industry. Williston, North Dakota is not shy about promoting the fact that they are located in a Right to Work state, offering workers the ability to opt out of union membership on an individual basis.

Another step to recruiting energy-intensive industries lies in preparing sites for their location. Large industrial facilities demand big sites with access to rail, roads, water, sewers, and cheap power. Failure to prepare these "job ready sites" will make recruitment of energy-intensive companies more challenging. Traditional utility companies can be partners in preparing sites for major industrial facilities. American Electric Power, a major Ohio based utility with customers in the Midwest and Southwest, created the AEP Quality Site program and identified 11 industrial sites that passed an initial evaluation and site visit.[69] Certified AEP Quality Sites have a minimum of 80 available and developable acres, are able to serve a range of industries, have utilities or a formal extension plan in place, have boundary surveys and topographic maps complete, and have environmental assessments and other due diligence completed.[70]

Build a Quality of Life

Small rural energy boom towns gain substantial new revenues from the initial stage of economic growth. Whether through impact fees, sales tax or other taxes, these communities gain millions in new revenue that can assist in not just developing the infrastructure needed to keep the energy growth happening, but also in making strategic investments in the region to build a higher quality of life. Quality of life factors matter to a younger generation of workers that have been leaving, not staying, in rural America.

Willison, North Dakota offers an example of quality of life improvements being generated out of new energy based government revenues. Built with new revenue from the local shale boom, Williston, North Dakota opened a $70,000,000 state of the art, community recreation center that offers everything from basketball courts to work out facilities to indoor surfing and golf.[71] A new sales tax that raised over $11,000,000 in the first 18 months helped launch this new recreation center.[72] Willison doubled in size to 30,000 residents, and they are making strategic investments to retain and attract a new generation of companies and workers.[73]

Amenities are an economic development tool to recruit and retain talent. Small, rural communities enjoying an energy boom need to capitalize on this growth by enacting new taxes on this industry to build recreational facilities that will last well beyond the initial energy boom.

Rural Connections through Technology

Small, rural energy boom communities need to build critical broadband connections as part of their long-term economic development strategy. Broadband is high-speed telecommunications services providing multiple channels of data over a single communications medium through either or both wireless or wireline facilities.[74] Broadband delivers everything from data packets on a wireless phone to massive digital transactions, on which many companies operate. Broadband supports rural economic development.[75] Rural broadband networks reduce the challenge and cost of distance by linking these isolated communities to larger urban and global markets through electronic means and act as a magnet for attracting new businesses.[76] In addition, younger workers and companies in general have simply come to expect a community to have broadband service.

Rural communities are far behind urban areas when it comes to the availability of broadband services. The cost of delivery of a broadband service is the main obstacle to rural America gaining this twenty-first century service. Rural communities cost just over $1,000 more per subscriber when compared to urban locations, due to the lack of rural population density.[77]

However, help is on the way. Programs such as the US Department of Agriculture's Broadband Loan Program get high marks for their economic impact.[78] As table 10.2 illustrates, the federal government offers many funding opportunities to develop rural broadband projects.

Table 10.2 Federal rural broadband funding programs

Agency	Program Title	Program Description
Federal Communications Commission	Healthcare Connect Fund	Broadband connectivity to health care providers and broadband networks
Federal Communications Commission	E-rate	$4,000,000,000 in discounted telecommunications, Internet access, and internal connections to schools and libraries
Department of Commerce	FirstNet	$7,000,000,000 for broadband networks
Department of Agriculture	Telecommunications Infrastructure Loan Program	Long-term direct and guaranteed loans for rural financing telecommunications and broadband service
Department of Agriculture	Rural Business Opportunity Grant	Grants rural community economic development, technology-based economic development, feasibility studies, leadership, entrepreneur training, business incubators, and strategic planning
Department of Agriculture	Rural Business Enterprise Grants	$10,000 to $500,000 grants for to finance and facilitate development of small and emerging businesses
Department of Agriculture	Rural Community Development Initiative	Grants for rural technical assistance and training for rural community and economic development
Department of Agriculture	Farm Bill Broadband Program	Loans for rural broadband projects
Department of Agriculture	Distance Learning and Telemedicine Program	Loans and grants for advanced telecommunications for learning and health care services
Department of Agriculture	Community Facility Program	Grants for small town, rural community facilities
Department of Agriculture	Community Connect	Funds for small town broadband infrastructure and computer community center

Source: Retrieved from http://www.commerce.wa.gov/Programs/Infrastructure/Broadband/Pages/Broadband-Funding-Sources.aspx.

Clearly, resources exist for rural communities to address broadband challenges. Small, rural energy boom towns may have additional revenue from local growth and development, and their long-term success demands that they focus on the development of a broadband network as part of needed long-term infrastructure investments.

Keep Energy Cheap

The cost of electricity and natural gas influences the successful operation of companies and the ability of states and regions to retain and attract jobs. The cost of energy is a top 10 corporate site location factor regularly cited by company executives considering a location decision. Often, for high-wage manufacturing and technology companies, the availability of reliable and low-cost power is the most important factor in a company's location decision.

The cost of power among the states varies. The explosion of shale gas drove down the price

of natural gas, but rates for electricity vary by region. States like Ohio rank twenty-eighth among the states for overall electricity costs, but as figure 10.1 illustrates, the availability of shale gas does not guarantee that overall energy costs will be lower.

Where a region is ranked impacts it ability to retain and attract companies. Utility companies provide power to customers and play a key role in economic development. A major decision for state policymakers is whether or not to regulate the power utility industry. The nation is nearly split evenly between states that regulate utilities like traditional monopolies or let the market determine who will provide power from a retail standpoint.

States regulating utilities as a monopoly have rates set by the government, typically on a guaranteed rate-of-return framework. Regulated utilities are guaranteed a set price and revenues based upon a commitment to

- North Dakota (9th)
- West Virginia (18th)
- Texas (23rd)
- Ohio (28th)
- Pennsylvania (34th)

Figure 10.1 Energy cost index shale state rankings.

Source: Small Business & Entrepreneurship Council 2012 Energy Index.

make a capital investment and an offer of service to anyone in their territory. Power companies build relationships with high-energy users in their territory, hoping they succeed and need more power.

States regulating utilities generally offer a "Rider Program" to reduce rates for high energy using companies, based upon power purchases and job commitments. As an example, the Duke Energy Carolinas offers an Economic Development Rider that grants a 4-year billing credit that can max out at 20 percent and declines to zero after 48 months. Customers are only eligible if they add a minimum of 1,000 kilowatts (kW) new load at one delivery point, a capital business investment of $400,000 per 1,000 kW of load added, plus a net increase of at least 75 full-time employees per 1,000 kW per new load in Duke Energy's service area. These customers agree to a minimum contract term with Duke Energy for 10 years and maintain a monthly average of 250 hours' usage of electric demand.

States such as Ohio, Pennsylvania, and Texas offer a competitively deregulated power market encouraging companies to market power rates at the retail level while regulating the production and distribution of power. Deregulated states offer few true incentives for utilities to be economic development partners. Instead, the competitive market is supposed to replace the role of the regulator and competition should promote lower prices and better service. As an example, the vast majority of Ohio customers have switched to a competitive company for their electricity. State leaders seeking to recruit energy-intensive companies can work to reduce power costs by:

1. Adopting either a regulatory or a competitive model for utility service;
2. Aggressively implementing a regulatory rate rider program for companies creating jobs; and
3. Embracing competition for power through a regulatory environment that is friendly to out of state competitors.

Build Energy Demand Side Programs

Small, rural energy boom towns benefit from the development of energy demand side programs that create incentives to organizations to use the new power source that is being extracted and processed. Pennsylvania's Natural Gas Vehicle Program offers an example of a successful energy demand side program. As part of the Pennsylvania's Act 13 several years ago, the Pennsylvania Natural Gas Vehicle Program was funded to help pay for the incremental purchase and conversion costs of heavy-duty natural gas fleet vehicles.[79] $6,000,000 was made available for the 2014–2015 fiscal year and awarded on a competitive basis for the project to cover no more than 50 percent of the incremental purchase or retrofit costs per vehicle, with a maximum of $25,000 per vehicle.[80] Funding is available for nonprofit organizations, for-profit companies, local transportation organizations, state owned or state related universities, Commonwealth or municipal authorities, and the Pennsylvania Turnpike Commission.[81] In 2013, 49 applications totaling $12,978,834 in grant requests with $1,397,292 from Local Transportation Organizations and $11,581,542 was requested for Non Local Transportation Organizations.[82] 19 projects were competitively selected, resulting in a total award of $6,809,263.[83] The 19 projects will result in:

- 302 CNG vehicles weighing greater than 14,000 lbs.;
- 27 LNG vehicles weighing greater than 14,000 lbs.;
- 3,700,000 gasoline gallon equivalents annually displaced; and
- 15 CNG and 1 LNG new Pennsylvania stations.[84]

States and regions looking to benefit from the growth of the shale industry need to capitalize on low-cost natural gas to transition vehicles to CNG or LNG, in order to reduce transportation costs.

Big Questions

1. Will the shale boom turn into a bust?
2. Can communities really stop the shale energy bust that is coming?
3. What are the short-term tactics that can lessen the negative impact of a shale energy bust for small rural communities?
4. What are the long-term tactics that can lessen the negative impact of a shale energy bust for small rural communities?
5. Are small, rural communities ready for a shale energy bust?

Class Writing Assignment

Remember, the Michigan Ag Association's proposed new tax on utilities to pay for a massive batch of biomass tax incentives to fund the largest development of utility scale, biomass power facilities in the United States. The utility industry is unexcited about the proposal, but the local communities and the biomass industry are supportive. The Michigan Ag Association believes their proposal will be part of a "bust prevention" strategy for rural communities in Michigan developing wind jobs and booming from the investment. The Michigan Ag Association has asked you to develop a quick plan to outline what five steps the state could take to avoid an economic "bust" with the wind industry—make sure to include the biomass tax incentives. They do not have a lot of money so they can only pay you to write two pages, double spaced, with your thoughts on how to avoid the bust that comes with a wind boom.

Class Debate

Remember, a Texas natural pipeline company plans to build the Wolverine Pipeline to take shale gas discovered in Ohio and ship it to a facility in rural Michigan. The Wolverine Pipeline is key to the state's recruitment of energy-intensive companies hoping low-cost energy will serve as a prime recruitment tool, and it matters to the recruitment of 200 with the chemical manufacturer Prime 101. Wolverine Pipeline received approval from the federal government for its construction; however, members of the Ohio General Assembly are not happy Michigan is gaining on the shale play. Representative Lee Smotch introduced legislation to abolish the natural gas industries authority to use eminent domain authority to purchase land. Proponents include the local communities who are developing the shale gas in Eastern Ohio, but the opponents are the entire business community, including the utilities, their unions, and chamber of commerce. Representative Smotch is Chairman of the Ohio House State and Local Government Committee, which will have a hearing on the bill and decide its fate. Representative Smotch will provide sponsor testimony, proponents and opponents will testify, and the committee will vote.

Additional Readings

2014 Energy Outlook, North America, Jones Lang Lasalle, Chicago, 2014.

Black, Dan, Terra McKinnish, and Set Sanders, "The Economic Impact of the Coal Boom and Bust," *The Economic Journal*, 115 (April 2005): 449–476.

Kelsey, Timothy, "Economic Implications of Natural Gas Drilling in the Marcellus Shale Region," Philadelphia Federal Reserve, *Cascade* 77 (Spring/Summer 2011).

Nolon, John, "Preventing the Ghost Town: What Rural Communities Need to Do to Survive in the Modern Economy," University of Kansas Law School, *Kansas Journal of Law and Policy Symposium*, February 21, 2014.

O'Leary, Sean and Ted Boettner, "Booms and Busts: The Impact of West Virginia's Energy Economy," West Virginia Center of Budget & Policy, July 2011.

Oyakawa, Javier, "Workforce Analysis for the Eagle Ford Shale," University of Texas at San Antonio, Center for Community and Business Research, October 2012.

Pender, John L., Jeremy G. Weber, and Jason P. Brown, "Sustainable Rural Development and Wealth Creation: Five Observations Based on Emerging Energy Opportunities," *Economic Development Quarterly*, 28, 1 (2014): 73–86.

Rutherford, Bruce, "Optimizing Economic Growth in the Shale Zone Communities," *Area Development Magazine*, 1st Quarter, 2014.

Severson, Kim, "Digital Age Is Slow to Arrive in Rural America," *New York Times*, February 17, 2011.

Notes

1 Energy Led Economic Development

1. http://www.eere.energy.gov/basics/renewable_energy/biomass.html.
2. National Renewable Energy Laboratory, Biomass Maps, retrieved from http://www.nrel.gov/gis/biomass.html.
3. US Department of Energy, Energy Information Agency, Hydro Power Fact Sheet.
4. "U.S. has large geothermal resources, but recent growth is slower than wind or solar," US Department of Energy, Today in Energy, November 18, 2011.
5. Ibid.
6. US Department of Energy, How Do Wind Turbines Work, retrieved from http://energy.gov/eere/wind/how-do-wind-turbines-work.
7. Ibid.
8. Ibid.
9. http://www.ohiocoal.com/faqs/.
10. Ibid.
11. US Environmental Protection Agency, Clean Energy Natural Gas Fact Sheet, retrieved from http://www.epa.gov/cleanenergy/energy-and-you/affect/natural-gas.html.
12. Ibid.
13. US Energy Information Administration, *Annual Energy Outlook 2010 with Projections to 2035* (Washington, DC: US Energy Information Administration, 2010).
14. Ibid.
15. US Environmental Protection Agency, Clean Energy Natural Gas Fact Sheet, retrieved from http://www.epa.gov/cleanenergy/energy-and-you/affect/natural-gas.html.
16. Joseph P. Tomain, "The Dominant Model of United States Energy Policy," 61 U. Colo. L. Rev. 363 (1990).
17. See Munn v. Illinois, 94 U.S. 113 (1876).
18. R. Richard Geddes, "A Historical Perspective on Electric Utility Regulation," *Cato Review of Business & Government, Regulation* (Winter 1992): 75–82.
19. Ibid.
20. Ibid.

21. Federal Power Act of 1935, Sec. 213, 49 Stat. 841–63 (1935).
22. Rural Electrification Act of 1936, 7 U.S. Code Sec. 31 (1936).
23. Energy Policy and Conservation Act of 1975, Public Law 94–163 (1975).
24. Power Plant and Industrial Fuel Use Act, Public Law 95–620 (1978).
25. See R. Bingham, E. W. Hill, and S. B. White, *Financing Economic Development: An Institutional Response* (Newbury Park, CA: Sage, 1990).
26. US Department of Energy, Energy Information Agency Report, 2014, retrieved from http://www.eia.gov/tools/faqs/faq.cfm?id=87&t=1.
27. Annual Average Electricity Price Comparison by State, State of Nebraska, 2011.
28. US Energy Information Agency, 2011 Annual Energy Review.
29. Raymond J. Keating, *Energy Cost Index 2012: Ranking the States* (Vienna, VA: Small Business and Entrepreneurship Council, 2012)
30. Ibid.
31. International Energy Agency, World Energy Outlook 2013 Factsheet, retrieved from http://www.worldenergyoutlook.org/media/weowebsite/factsheets/WEO2013_Factsheets.pdf.
32. Ibid.
33. Ibid.
34. Ibid.
35. *How to Conduct an Energy Audit: A Short Guide for Local Governments and Communities* (Trenton, NJ: New Jersey Department of Environmental Protection Division of Science, Research and Technology Bureau of Sustainable Communities and Innovative Technologies, 2006).
36. Ibid.
37. Ibid.
38. US Environmental Protection Agency, Clean Energy Report, Chapter 6, Energy Efficiency Best Practices Report, 2014, retrieved from http://www.epa.gov/cleanenergy/documents/suca/napee_chap6.pdf.
39. Ibid.
40. Ibid.
41. Database for State Renewable and Efficiency, retrieved from http://www.dsireusa.org/glossary/.
42. Ibid.
43. *15 Economic Facts about Millennials* (Washington, DC: The White House, The Council of Economic Advisers, October 2014).
44. Fred Dews, "11 Facts about the Millennial Generation," Brookings Institute, June 2, 2014, retrieved from http://www.brookings.edu/blogs/brookings-now/posts/2014/06/11-facts-about-the-millennial-generation.
45. Tony Dutzik, Jeff Inglis, and Phineas Baxandall, *Millennials in Motion: Changing Travel Habits of Young Americans and the Implications for Public Policy* (Boston: US PIRG Education Fund & Frontier Group, 2014).
46. Retrieved from http://www.nielsen.com/content/corporate/us/en/insights/news/2014/millennials-prefer-cities-to-suburbs-subways-to-driveways.html.
47. Katherine Peralta, "Homeownership for Millennials Declines to New Lows New Data Show Slight Drop in Homeownership for Americans under 35," *U.S. News & World Report*, April 30, 2014.

48. Bana Jobe, "6 Ways Millennials Will Change the Energy Industry from the Inside Out," September 4, 2014, http://thewire.oncor.com/News/Pages/6-Ways-Millennials-Will-Change-the-Energy-Industry-from-the-Inside-Out.aspx.
49. Ibid.
50. *Global Metro Monitoring* (Washington, DC: Brookings Institution, 2011–2012).
51. Ibid.
52. Mark Muro, Jonathan Rothwell, and Devashree Saha, *Sizing the Clean Economy: A National and Regional Green Jobs Assessment* (Washington, DC: Brookings Institution, July 13, 2011).
53. Retrieved from http://www.historydetroit.com/part_2.asp.
54. Retrieved from http://www.detroit.com/history/.
55. Retrieved from http://www.somacon.com/p469.php.
56. Retrieved from http://news.bbc.co.uk/2/hi/business/6346299.stm.
57. Ibid.
58. Retrieved from http://www.67riots.rutgers.edu/d_index.htm.
59. See 2010 Milken Institute State Science and Technology Index, http://www.milkeninstitute.org/tech/tech2010.taf.
60. Ibid.
61. See Michigan Still Struggles on Entrepreneurship Score Card, CBS Detroit, March 6, 2011, retrieved from http://detroit.cbslocal.com/2011/03/06/michigan-still-struggles-on-entrepreneurship-score-card/.
62. Ibid.
63. Ibid.
64. *Michigan 2010–2011*, Michigan Small Business Foundation Score Card, *Small Business Association of Michigan*, p. 9 retrieved from https://www.sbam.org/Resources/newsid438/393/2011-Entrepreneurship-Score-Card-Shows-Michigan-Holding-Steady.
65. Ibid.
66. Ibid.
67. Retrieved from http://www.census.gov/popest/metro/CBSA-est2009-annual.html.
68. Ron French, Leonard N. Fleming, and Mike Wilkinson, "Detroit's Population Falls 25%, Bing Wants Recount," *The Detroit News*, March 23, 2011.
69. Ibid.
70. Retrieved from http://www.lasvegasnevada.gov/FactsStatistics/history.htm.
71. Ibid.
72. Ibid.
73. Retrieved from http://www.lasvegasnevada.gov/FactsStatistics/history.htm.
74. Retrieved from http://inventors.about.com/library/weekly/aa081797.htm.
75. Retrieved from http://www.lasvegasnevada.gov/FactsStatistics/history.htm.
76. Retrieved from www.americangaming.org/Industry/factsheets/statistics_detail.cfv?id=4.
77. Retrieved from http://www.lasvegassun.com/news/2010/jun/11/building-boom-recessionary-bust-local-real-estate-/.
78. Ibid.
79. Ibid.

80. Retrieved from http://www.bls.gov/ro9/unionnv.htm.
81. Ibid.
82. Retrieved from http://www.lasvegassun.com/news/2010/jun/11/building-boom-recessionary-bust-local-real-estate-/.
83. Ibid.
84. Ibid.
85. Ibid.
86. Ibid.
87. Ibid.
88. Retrieved from http://www.americangaming.org/assets/files/studies/States_of_the_States_2011_FINAL.PDF.
89. Ibid.
90. Ibid.
91. Retrieved from http://gaming.unlv.edu/abstract/macau.html.
92. www.lvrj.com/news/47510187.html.
93. Ibid.
94. Ibid.
95. http://www.nevadabusiness.com/issue/1009/27/2119.
96. http://www.ceosforcities.org/pagefiles/City_Dividends_FINAL_NewLook.pdf.
97. See http://www.eda.gov/Research/ClusterBased.xml.
98. Ibid.
99. Michael E. Porter, "Location, Competition and Economic Development: Local Clusters in a Global Economy," *Economic Development Quarterly*, 14, 1 (2000): 15–34.
100. See *Using Employment Data to Better Understand Your Local Economy, Tool 4. Shift-Share Analysis Helps Identify Local Growth Engines* (Penn State University, 2013), retrieved from http://pubs.cas.psu.edu/freepubs/pdfs/UA373.pdf.
101. Ibid.

2 Economic Development and Energy Policy

1. John S. Gordon, *An Empire of Wealth: The Epic History of American Economic Power* (New York: HarperCollins, 2004), p. 35.
2. www.http://xroads.virginia.edu/~cap/citybeautiful.
3. Ibid.
4. Ibid.
5. Ibid.
6. Ibid.
7. Ibid.
8. Ibid., p. 10.

9. Ibid., p. 11.
10. Ibid.
11. Ibid.
12. See Adam Cohen and Elizabeth Taylor, *American Pharaoh: Mayor Richard J. Daley—His Battle for Chicago and the Nation* (Boston: Little, Brown, 2001).
13. Gordon, *An Empire of Wealth*, p. 296.
14. Ibid.
15. Ibid., p. 298.
16. Ibid.
17. Ibid.
18. Ibid., p. 12.
19. Ibid., p. 13.
20. Ibid., p. 17.
21. Ibid.
22. Ibid.
23. See The Global City: New York, London and Tokyo, Saskia Sassen, (Princeton University Press, Princeton, NJ, 2001).
24. IEDC, p. 19.
25. Ibid.
26. Ibid.
27. Ibid., p. 20.
28. Ibid.
29. Ibid., p. 21.
30. Ibid.
31. Ibid., p. 22.
32. Ibid.
33. Joseph P. Tomain, "The Dominant Model of United States Energy Policy," *61 U. Colo. L. Rev.* 363, 355–392 (1990).
34. Ibid.
35. See Munn v. Illinois, 94 U.S. 113 (1876).
36. R. Richard Geddes, "A Historical Perspective on Electric Utility Regulation," *Cato Review of Business & Government, Regulation* (Winter 1992): 75–82.
37. Ibid.
38. Ibid.
39. Tomain, "The Dominant Model of United States Energy Policy."
40. Ibid.
41. Ibid.
42. Ibid.
43. Ibid.
44. Ibid.
45. Federal Water Power Act of 1920, Chapter 285, 41 Stat. 1063 (1920).
46. See generally, N. Ely, "The Conservation of Oil," *51 Harvard L. Rev.* 1209 (1938) 1142–1151.
47. Ibid.
48. Ibid.
49. Tomain, "The Dominant Model of United States Energy Policy."

50. Ibid.
51. Ibid.
52. Federal Power Act of 1935, Sec. 213, 49 Stat. 841–63 (1935).
53. Ibid.
54. Rural Electrification Act of 1936, 7 U.S. Code Sec. 31 (1936).
55. Natural Gas Act of 1938, 15 U.S. Code Sec. 717 (1938).
56. Tomain, "The Dominant Model of United States Energy Policy."
57. Ibid.
58. Ibid.
59. Atomic Energy Act of 1946, Public Law 585, 79th Congress (1946).
60. Ibid.
61. Atomic Energy Act of 1954.
62. Energy Reorganization Act of 1974, 42 U.S. Code Sec. 1 (1974).
63. Energy Policy and Conservation Act of 1975, Public Law 94–163 (1975).
64. Department of Energy Organizational Act of 1975, Public Law 95–91 (1975).
65. See generally, Public Utility Regulatory Policies Act, Public Law 95–617 (1978), Energy Tax Act, Public Law 95–618 (1978), and Natural Gas Policy Act, Public Law 95–621 (1978).
66. National Energy Conservation Policy Act, Public Law 95–619 (1978).
67. Ibid.
68. Power Plant and Industrial Fuel Use Act, Public Law 95–620 (1978).
69. Public Utility Regulatory Policies Act, Public Law 95–617 (1978).
70. Energy Security Act of 1980, Public Law 96–294 (1980).
71. Ibid.
72. Ibid.
73. Ibid.
74. Energy Policy Act of 1992, Public Law 102–486.
75. Ibid.
76. Ibid.
77. Tomain, "The Dominant Model of United States Energy Policy."
78. Retrieved from http://www.diversifynevada.com/key-industries/energy.
79. Ibid.
80. See http://www.lasvegassun.com/news/2011/sep/27/interest-residing-downtown-soars/.
81. Retrieved from http://www.businessinsider.com/r-tesla-factory-near-reno-is-bet-old-nevada-will-meet-the-new--2014-10.
82. http://www.nytimes.com/2014/09/13/business/nevada-woos-tesla-plant-in-tax-deal-but-economic-benefits-prompt-debate.html?_r=0.
83. Retrieved from http://www.seia.org/state-solar-policy/nevada.
84. Ibid.
85. Ibid.
86. Ibid.
87. Ibid.
88. Comprehensive Economic Development Strategy Las Vega Global Economic Alliance, TIP Strategies, Inc., July 3, 2013, retrieved from http://www.lasvegasnevada.gov/files/Southern_NV_CEDS.pdf.

89. Retrieved from http://www.reviewjournal.com/business/economy/amonix-closes-north-las-vegas-solar-plant-after-14-months.
90. Frank Witsal, "Michigan's Wind Energy Industry Soaring," *Detroit Free Press*, July 1, 2014, retrieved from http://archive.freep.com/article/20140629/BUSINESS06/306290068/wind-alternative-energy-michigan-epa.
91. *Clean Economy Rising, Wind Powers Michigan's Energy Industry* (Washington, DC: Pew Center, 2014), retrieved from http://www.pewtrusts.org/~/media/Assets/2014/10/MichiganPolicyBrief_v4_web.pdf?la=en.
92. Ibid.
93. Ibid.
94. Ibid.

3 Energy Supply and Demand

1. US Energy Information Agency, retrieved from http://www.eia.gov/energyexplained/index.cfm?page=about_home.
2. US Energy Information Agency, Annual Energy Outlook 2013.
3. Ibid.
4. Ibid.
5. Ibid.
6. Ibid.
7. Ibid.
8. Ibid.
9. International Energy Agency, World Energy Outlook 2013 Factsheet.
10. US Department of Energy, Energy Information Agency, Annual Electric Report.
11. Ibid.
12. Ibid.
13. http://www.eere.energy.gov/basics/renewable_energy/biomass.html.
14. Ibid.
15. Rik Stevens, "Mid-December Opening Set for Biomass Plant in Berlin," *Concord Monitor*, November 15, 2013.
16. Ibid.
17. Ibid.
18. Ibid.
19. Ibid.
20. National Renewable Energy Laboratory, Biomass Maps, retrieved from http://www.nrel.gov/gis/biomass.html.
21. Ibid.
22. Ibid.
23. Union of Concerned Scientist, Biomass Fact Sheet, retrieved from http://www.ucsusa.org/clean_vehicles/smart-transportation-solutions/cleaner_fuels/ethanol-and-other-biofuels/biomass-energy-resources.html.
24. Ibid.

25. Ibid.
26. Ibid.
27. Ibid.
28. Ibid.
29. Ibid.
30. See Jim Carlton, "(Bio)Mass Confusion, High Costs and Environmental Concerns Have Pushed Biomass Power to the Sidelines in the US," *Wall Street Journal*, October 18, 2010, retrieved from http://www.wsj.com/articles/SB10001424052748704116004575521713188253290.
31. http://www.ohiocoal.com/faqs/.
32. US Energy Department, Energy Information Agency, Energy in Brief Coal Report, 2014.
33. Ibid.
34. Ibid.
35. Ibid.
36. Ibid.
37. Jonathan M. Roenker, *The Economic Impact of Coal in Appalachian Kentucky* (Lexington, KY: Center for Business and Economic Research, University of Kentucky, 2001.
38. US Energy Department, Energy Information Agency, Coal Report, 2014.
39. Ibid.
40. US Energy Department, Energy Information Agency, Energy in Brief Coal Report, 2014.
41. US Energy Department, Energy Information Agency, Coal Report, 2014.
42. Ibid.
43. US Energy Department, Energy Information Agency, Energy in Brief Coal Report, 2014.
44. Ibid.
45. Ibid.
46. Ibid.
47. Ibid.
48. Ibid.
49. http://www1.eere.energy.gov/geothermal/geothermal_basics.html.
50. Ibid.
51. John A. Shonder, Michaela A. Martin, and Patrick J. Hughes, *Geothermal Heat Pumps in K-12 Schools: A Case Study of the Lincoln, Nebraska Schools* (Oak Ridge, TN: Oak Ridge National Laboratory, ORNL/TM-2000/80, February 4, 2013), p. 12, retrieved from http://www.ornl.gov/sci/ees/etsd/btric/pdfs/com_ghpsinschools.pdf.
52. Oak Ridge National Laboratory, *Big Savings from the World's Largest Installation of Geothermal Heat Pumps at Fort Polk, Louisiana* (Oak Ridge, TN: Department of Energy, Oak Ridge National Laboratory, February 5, 2013), retrieved from http://www.ornl.gov/sci/femp/pdfs/fortpolk.pdf.
53. "U.S. Has Large Geothermal Resources, but Recent Growth Is Slower than Wind or Solar," US Department of Energy, Today in Energy, November 18, 2011.
54. Ibid.

55. Ibid.
56. Ibid.
57. US Department of Energy, Energy Information Agency, Hydro Power Fact Sheet.
58. "Maine Project Takes Historic Step Forward in U.S. Tidal Energy Deployment," US Department of Energy, News Report, May 4, 2012, retrieved from http://energy.gov/articles/maine-project-takes-historic-step-forward-us-tidal-energy-deployment.
59. Ibid.
60. Ibid.
61. Ibid.
62. Ibid.
63. US Department of Energy, Energy Information Agency, Hydro Power Fact Sheet.
64. Ibid.
65. Ibid.
66. "Top 10 Things You Didn't Know about Hydropower," US Department of Energy, Energy Information Agency, April 19, 2013, retrieved from http://energy.gov/articles/top-10-things-you-didnt-know-about-hydropower.
67. Boualem Hadjerious, "An Assessment of Energy Potential at Non-Powered in the United States," US Department of Energy, April 2012.
68. Ibid.
69. Kate Galbraith, "Retrofitting Dams to Generate Electricity," *New York Times*, August 19, 2009.
70. "How Nuclear Power Works," Union of Concerned Scientist, retrieved from http://www.ucsusa.org/nuclear_power/nuclear_power_technology/how-nuclear-power-works.html.
71. Kevin Bullis, "Can Small Reactors Ignite a Nuclear Renaissance?" *MIT Technology Review*, March 28, 2013.
72. Ibid.
73. "How Nuclear Power Works," Union of Concerned Scientist, retrieved from http://www.ucsusa.org/nuclear_power/nuclear_power_technology/how-nuclear-power-works.html.
74. Ibid.
75. Ibid.
76. Ibid.
77. Ibid.
78. Ibid.
79. Ibid.
80. Ibid.
81. US Department of Energy, Energy Information Agency, Nuclear Fact Sheet.
82. See http://www.api.org/aboutoilgas/.
83. Ibid.
84. Retrieved from http://science.howstuffworks.com/environmental/energy/oil-refining2.htm.
85. Ibid.

86. Ibid.
87. US Environmental Protection Agency, Clean Energy Natural Gas Fact Sheet, retrieved from http://www.epa.gov/cleanenergy/energy-and-you/affect/natural-gas.html.
88. Ibid.
89. "What Is Shale Gas and Why Is It Important?" US Department of Energy, Energy Information Agency, Energy in Brief, December 5, 2012.
90. Ibid.
91. Ibid.
92. Ibid.
93. US Environmental Protection Agency, Region 1, Solar Energy Fact Sheet, retrieved from http://www.epa.gov/region1/eco/energy/re_solar.html.
94. Ibid.
95. Ibid.
96. Ibid.
97. Ibid.
98. US Department of Energy, Energy Information Agency, Solar Fact Sheet, retrieved from http://www.eia.gov/kids/energy.cfm?page=solar_home-basics-k.cfm.
99. US Environmental Protection Agency, Region 1, Solar Energy Fact Sheet, retrieved from http://www.epa.gov/region1/eco/energy/re_solar.html.
100. Ibid.
101. US Department of Energy, Energy Information Agency, Solar Fact Sheet, retrieved from http://www.eia.gov/kids/energy.cfm?page=solar_home-basics-k.cfm.
102. Brad Plumer, "These Maps Show the Best Places to Put Solar and Wind Power. (It's Not Where You Think.)," *Washington Post*, July 15, 2013.
103. Retrieved from http://www.acciona-energia.com/activity_areas/csp/installations/nevadasolarone/nevada-solar-one.aspx.
104. Ibid.
105. Ibid.
106. Ibid.
107. "Nevada Continues Strong Solar Investment,"Solar Energy Industries Association, Friday, December 19, 2014, retrieved from http://www.seia.org/news/nevada-continues-strong-solar-investment.
108. US Department of Energy, Annual Energy Outlook 2014.
109. Ibid.
110. US Department of Energy, How Do Wind Turbines Work, retrieved from http://energy.gov/eere/wind/how-do-wind-turbines-work.
111. Ibid.
112. Ibid.
113. Ibid.
114. Ibid.
115. Ibid.
116. Retrieved from http://www.awea.org/learnabout/publications/factsheets/factsheets_windenergybasics.cfm.
117. Ibid.

118. Ibid.
119. Ibid.
120. Ibid.
121. *2012 Wind Technologies Market Report*, US Department of Energy, August 6, 2013.
122. Ibid.
123. Ibid.
124. Retrieved from http://www.awea.org/learnabout/publications/factsheets/factsheets_windenergybasics.cfm.
125. *2012 Wind Technologies Market Report*, US Department of Energy, August 6, 2013.
126. Ibid.
127. Ibid.
128. Ibid.
129. Ibid.
130. Ibid.
131. Retrieved from http://science.time.com/2013/08/14/power-drain-the-digital-cloud-is-using-more-energy-than-you-think/.
132. Ibid.
133. Ibid.
134. US Energy Information Agency, Annual Energy Outlook 2013.
135. Ibid.
136. Ibid.
137. Ibid.
138. Ibid.
139. Ibid.
140. Ibid.
141. Ibid.
142. Ibid.
143. International Energy Agency, World Energy Outlook 2013 Factsheet.
144. Ibid.
145. Ibid.
146. Ibid.
147. Ibid.
148. Ibid.
149. Ibid.
150. Ibid.
151. Ibid.
152. See www.DiversifyNevada.com.
153. Retrieved from http://energy.nv.gov/Programs/Green_Building_(LEED)_Tax_Abatements/.
154. Retrieved from http://dsireusa.org/incentives/incentive.cfm?Incentive_Code=NV01F&re=0&ee=0.
155. Ibid.
156. Retrieved from http://dsireusa.org/incentives/incentive.cfm?Incentive_Code=NV22F&re=0&ee=0.

157. Ibid.
158. Retrieved from http://energy.nv.gov/Programs/Revolving_Loans_for_Renewable_Energy/.
159. Retrieved from http://energy.nv.gov/Programs/State_Energy_Program_Formula_Grant/.
160. Retrieved from http://dsireusa.org/incentives/incentive.cfm?Incentive_Code=NV23F&re=0&ee=0.
161. Ibid.
162. Retrieved from http://dsireusa.org/incentives/incentive.cfm?Incentive_Code=NV09F&re=0&ee=0.
163. Retrieved from http://www.dri.edu/about.
164. Ibid.
165. *A Special Message from Governor Rick Snyder, Ensuring Our Future: Energy and the Environment*, November 28, 2012, retrieved from http://www.michigan.gov/documents/snyder/EE_Message_FINAL_pdf_404563_7.pdf.
166. Ibid.
167. Retrieved from http://www.dsireusa.org/incentives/allsummaries.cfm?State=MI&&re=0&ee=0.
168. Ibid.
169. Ibid.
170. Ibid.
171. Ibid.
172. Ibid.
173. Ibid.
174. Ibid.
175. Ibid.
176. Ibid.
177. Ibid.
178. Ibid.
179. Ibid.
180. Ibid.
181. Ibid.
182. Ibid.
183. Ibid.
184. Ibid.
185. Ibid.
186. Ibid.

4 Energy Led Economic Development Case Study

1. Retrieved from http://www.laits.utexas.edu/gov310/PEP/policy/.
2. Ibid.

3. Ibid.
4. Ibid.
5. Ibid.
6. See Dietram A. Scheufele and David Tewksbury, "Framing, Agenda Setting, and Priming: The Evolution of Three Media Effects Models," *Journal of Communication*, 57, 1 (2007): 9–20.
7. Ohio Oil and Gas Association Fact Sheet, retrieved from http://208.68.184.222/wp-content/uploads/2011/12/Economic-Impact-Study-Results-2011.pdf.
8. Lesley Hunter, "Renewable Energy in the 50 States: Midwestern Region," American Council on Renewable Energy, 2013.
9. Ibid.
10. Ibid.
11. Ibid.

5 Energy Conservation and Efficiency

1. US Environmental Protection Agency, Clean Energy Report, Chapter 6, Energy Efficiency Best Practices Report, 2014, retrieved from http://www.epa.gov/cleanenergy/documents/suca/napee_chap6.pdf.
2. Ibid.
3. Ibid.
4. Ibid.
5. Kenneith Gillingham, Richard G. Newell, and Karen Palmer, "Energy Efficiency Economics and Policy," *Annual Review of Resource Economics*, NBER Working Paper No. w15031, June 2009, retrieved from http://papers.ssrn.com/sol3/papers.cfm?abstract_id=1413595.
6. Ibid.
7. Ibid.
8. Ibid.
9. Ibid.
10. Ibid.
11. American Council for an Energy Efficient Economy, Economic Development Fact Sheet, September 2012, retrieved from www.aceee.org/blog/2012/09/energy-efficiency-and-economic-opportunity.
12. Ibid.
13. US Department of Energy, Energy Information Agency, American Energy Outlook, McKinsey Analysis, 2008.
14. Ibid.
15. Jamie Howland and Derek Murrow, "Energy Efficiency: Engine of Economic Growth," ENE Environment Northeast, October 2009.
16. Ibid.
17. Ibid.
18. Ibid.

19. David Frankel and Humayun Tai, "Giving US Energy Efficiency a Jolt," McKinsey, December 2013, retrieved from http://www.mckinsey.com/insights/energy_resources_materials/giving_us_energy_efficiency_a_jolt.
20. Ibid.
21. Ibid.
22. Ibid.
23. Ibid.
24. Ibid.
25. US Department of Energy, Energy Information Agency, American Energy Outlook, McKinsey Analysis, 2008.
26. Ibid.
27. See generally, Jacques-Francois Thisse, "Location Theory, Regional Science and Economics," *Journal of Regional Science*, 27 (1987): 519–528.
28. National Federation of Independent Business, 2014, retrieved from http://www.nfib.com/advocacy/energy/.
29. Mali R. Schantz-Feld, "Top Site Selection Factors: Energy Availability and Costs—Reconciling Price and Reliability," *Area Development*, November 2011, retrieved from http://www.areadevelopment.com/EnergyEnvironment/November2011/site-selection-factors-energy-costs-88872252.shtml.
30. US Energy Information Agency Electricity Use Report, 2011, retrieved from http://www.eia.gov/energyexplained/index.cfm?page=electricity_use.
31. US Department of Energy, Energy Information Agency Report, 2014, retrieved from http://www.eia.gov/tools/faqs/faq.cfm?id=87&t=1.
32. US Department of Energy, Energy Information Agency Report, 2014, retrieved from http://www.eia.gov/tools/faqs/faq.cfm?id=85&t=1.
33. Ibid.
34. Ibid.
35. US Department of Energy, Energy Information Agency, State Profiles, 2014, retrieved from http://www.eia.gov/state/?sid=US.
36. Statemaster.com, 2014, retrieved from http://www.statemaster.com/graph/ene_tot_ele_con-energy-total-electricity-consumption.
37. American Council for an Energy Efficient Economy, retrieved from http://www.aceee.org/.
38. Ibid.
39. Ibid.
40. Ibid.
41. Ibid.
42. Ibid.
43. Ibid.
44. Ibid.
45. Ibid.
46. Ibid.
47. Ibid.
48. US Department of Energy, Energy Information Agency, Energy Consumption Report, 2014, retrieved from http://www.eia.gov/energyexplained/index.cfm?page=us_energy_use.

49. US Department of Energy, Manufacturing Energy Consumption Survey, 2013, retrieved from http://www.eia.gov/consumption/manufacturing/reports/2010/decrease_use.cfm?src=Consumption-f4.
50. Ibid.
51. Ibid.
52. Ibid.
53. Ibid.
54. *Global Industrial Energy Efficiency Benchmarking: An Energy Policy Tool*, United Nations Industrial Development Organization, Working Paper, November 2010, retrieved from http://unido.org/fileadmin/user_media/Services/Energy_and_Climate_Change/Energy_Efficiency/Benchmarking_%20Energy_%20Policy_Tool.pdf.
55. Ibid.
56. Retrieved from http://www.epa.gov/statelocalclimate/local/topics/energy-efficiency.html.
57. US Department of Energy, Residential Energy Consumption Survey, 2009.
58. Ibid.
59. Ibid.
60. Ibid.
61. Ibid.
62. US Department of Energy, Energy Information Agency, Residential Energy Consumption Survey, 2012, retrieved from http://www.eia.gov/todayinenergy/detail.cfm?id=10271&src=‹ Consumption Residential Energy Consumption Survey (RECS)-f1.
63. Ibid.
64. Ibid.
65. Ibid.
66. Ibid.
67. Ibid.
68. Ibid.
69. World Bank Report, Motor Vehicle Use per 1000 Vehicles, retrieved from http://data.worldbank.org/indicator/IS.VEH.NVEH.P3.
70. Ibid.
71. Ibid.
72. US Department of Energy, Energy Information Agency, International Energy Outlook, 2013, retrieved from http://www.eia.gov/forecasts/ieo/transportation.cfm.
73. Ibid.
74. Ibid.
75. Ibid.
76. US Department of Energy, Energy Information Agency, Energy in Brief, 2014, retrieved from http://www.eia.gov/energy_in_brief/article/major_energy_sources_and_users.cfm.
77. US Department of Energy, *Transportation Energy Data Book*, Table 2.5, 2011, retrieved from http://cta.ornl.gov/data/chapter2.shtml.
78. Retrieved from http://www.energystar.gov/.

79. Ibid.
80. Ibid.
81. Ibid.
82. Ibid.
83. Ibid.
84. Ibid.
85. Ibid.
86. Ibid.
87. Ibid.
88. *How to Conduct an Energy Audit: A Short Guide for Local Governments and Communities* (Trenton,, New Jersey Department of Environmental Protection Division of Science, Research and Technology Bureau of Sustainable Communities and Innovative Technologies, 2006).
89. Ibid.
90. Ibid.
91. Ibid.
92. Ibid.
93. Ibid.
94. Ibid.
95. Ibid.
96. Ibid.
97. US Department of Energy Report, Energy Audits, retrieved from http://www.nrel.gov/docs/fy11osti/50365.pdf.
98. Ibid.
99. Ibid.
100. Ibid.
101. Juita-Elena Yusuf and Kathaine A. Neill, "State Energy-Based Economic Development Policies and Examples," *Economic Development Quarterly*, 27, 3 (2013): 246–247.
102. Database for State Renewable and Efficiency, retrieved from http://www.dsireusa.org/glossary/.
103. Retrieved from http://energy.gov/savings.
104. Database for State Renewable and Efficiency, retrieved from http://www.dsireusa.org/glossary/.
105. Ibid.
106. Ibid.
107. Ibid.
108. Ibid.
109. Ibid.
110. Ibid.
111. Ibid.
112. Ibid.
113. Ibid.
114. Ibid.
115. Ibid.
116. Ibid.

117. Ibid.
118. Ibid.
119. Ibid.
120. US Environmental Protection Agency, Clean Energy Report, Chapter 6, Energy Efficiency Best Practices Report, 2014, retrieved from http://www.epa.gov/cleanenergy/documents/suca/napee_chap6.pdf.
121. Ibid.
122. Ibid.
123. Database for State Renewable and Efficiency, retrieved from http://www.dsireusa.org/glossary/.
124. Ibid.
125. Ibid.
126. Ibid.
127. Ibid.
128. Ibid.
129. Ibid.
130. Ibid.
131. Ibid.
132. Ibid.
133. Ibid.
134. Ibid.
135. Ibid.
136. Ibid.
137. US Environmental Protection Agency, Clean Energy Report, Chapter 6, Energy Efficiency Best Practices Report, 2014, retrieved from http://www.epa.gov/cleanenergy/documents/suca/napee_chap6.pdf.
138. Ibid.
139. Database for State Renewable and Efficiency, retrieved from http://www.dsireusa.org/glossary/.
140. Ibid.

6 Energy as Critical Infrastructure

1. Dan Weikel, "San Bruno Pipeline Explosion: 'A Failure of the Entire System,'" *Los Angeles Times*, August 30, 2011.
2. Ibid.
3. Ibid.
4. "PG&E To Offer Targeted Electric Rate Reductions to Promote Economic Development ","* The Business Journal*, October 3, 2013, retrieved from http://www.bizjournals.com/sacramento/prnewswire/press_releases/California/2013/10/03/SF91784.
5. Ibid.
6. Ibid.

7. Ibid.
8. International Energy Agency, World Energy Outlook 2013 Factsheet.
9. http://www.epa.gov/ogd/forms/Definition_of_Infrastructure_for_ARRA.pdf.
10. Kevin T. Deno, "The Effect of Public Capital on U.S. Manufacturing Activity: 1970–78," *Southern Economic Journal*, 55 (October, 1988): 400–411.
11. See Brian Berry and John Parr, *Market Centers and Retail Location Theory and Applications* (Engelwood Cliffs, NJ: Prentice Hall, 1988).
12. See Alicia Munnell, "How Does Public Infrastructure Affect Regional Economic Performance," *New England Economic Review* (September/October 1990): 25, 11–32.
13. US Energy Information Agency.
14. American Society of Civil Engineers, 2013 Infrastructure Report Card.
15. Ibid.
16. Ibid.
17. Ibid.
18. Twenty-fifth Annual Corporate Survey, The economy's gradual recovery during 2010 has resulted in increased business optimism, a rise in new facility plans, and some changes in site selection priorities, Geraldine Gambale, Ed., *Area Development Magazine* (Winter 2011), retrieved from http://www.areadevelopment.com/AnnualReports/jan2011/25th-annual-corporate-executive-survey48843.shtml.
19. Ibid.
20. US Energy Information Agency, 2011 Annual Energy Review (Washington, D.C.) 2011, retrieved from http://www.eia.gov/totalenergy/data/annual/pdf/aer.pdf.
21. Ibid.
22. Raymond J. Keating, *Energy Cost Index 2012: Ranking the States* (Vienna, VA: Small Business and Entrepreneurship Council, 2012.
23. Annual Average Electricity Price Comparison by State, State of Nebraska, 2011.
24. Keating, "Energy Cost Index 2012."
25. Ibid.
26. US Energy Information Agency, 2013 International Energy Report, retrieved from http://www.eia.gov/forecasts/ieo/industrial.cfm.
27. Ibid.
28. Ibid.
29. Ibid.
30. Ibid.
31. American Chemical Council Economic Impact Report, 2013, retrieved from http://www.americanchemistry.com/Jobs/EconomicStatistics.
32. See http://www.payscale.com/research/US/Industry=Chemical_Manufacturing/Salary.
33. James Manyika, Michael Chui, Brad Brown, Jacques Bughin, Richard Dobbs, Charles Roxburgh, and Angela Hung Byers, *Big Data: The Next*

Frontier for Innovation, Competition and Productivity (McKinsey Global Institute, 2011) Washington, DC.
34. Ibid.
35. Ibid.
36. Anne Fisher, "Big Data Could Generate Millions of New Jobs," *Fortune Magazine*, May 21, 2013.
37. See http://www.businessinsider.com/10-tech-skills-that-will-instantly-net-you-100000-salary-2012-8?op=1.
38. Retrieved from http://www.blsstrategies.com/Docs/Events/Event_33.pdf.
39. Ibid.
40. Retrieved from http://www.datacenterknowledge.com/archives/2013/04/29/the-billion-dollar-data-centers/.
41. Ibid.
42. Ibid.
43. Retrieved from http://www.blsstrategies.com/Docs/Events/Event_33.pdf.
44. Ibid.
45. Retrieved from http://yosemite.epa.gov/opa/admpress.nsf/0/810cb77be1d695a58525773b004ed344.
46. TVA Fact Sheet, retrieved from http://www.tva.com/abouttva/history.htm.
47. Ibid.
48. Ibid.
49. Ibid.
50. Ibid.
51. Ibid.
52. "TVA Touts Economic Development Impact Utility Says Its Activities Added 10,750 Jobs in '12," *Knoxville News Sentinel*, November 15, 2012, retrieved from http://www.knoxnews.com/news/2012/nov/15/tva-touts-economic-development-impact/.
53. Ibid.
54. TVA Fact Sheet, retrieved from http://www.tva.com/news/keytopics/power_prices.htm.
55. Roberto F. Aguilera and Marian Radetzki, "Shale Gas and Oil: Fundamentally Changing Global Energy Markets," *Oil and Gas Journal*, 111, 12 (December 2, 2013): 54–61.
56. US Energy Information Agency, 2012 Annual Energy Outlook.
57. "Shale Oil: The Next Energy Revolution," *PriceWaterhouseCoopers*, February 2013.
58. Retrieved from http://www.electricitybIbid.com/electricity/index.php/2008/05/05/list-of-electricity-deregulated-states-in-the-usa/.
59. Retrieved from http://locationindiana.com/about-duke-energy/power-quality.asp.
60. Ibid.
61. Ibid.
62. Retrieved from http://www.duke-energy.com/pdfs/DE-IN_Rider_58_05_07_13.pdf.
63. Ibid.

64. Ibid.
65. Ibid.
66. Ibid.
67. Retrieved from http://www.duke-energy.com/pdfs/DE-IN_Rider_54_05_07_13.pdf.
68. Ibid.
69. Robert McClanahan, "Electric Deregulation," *IEEE Industry Applications Magazine*, March–April, 2002.
70. Ibid.
71. Ibid.
72. Ibid.
73. Ibid.
74. Ibid.
75. Annual Average Electricity Price Comparison by State, State of Nebraska, 2011.
76. James Osbource, "Report: Texas' Deregulated Power Prices below National Average but Still Too High," *Dallas Morning News*, December 19, 2013. http://www.dallasnews.com/business/energy/20131219-report-texas-deregulated-power-prices-below-national-average-but-still-too-high.ece
77. Retrieved from http://www.energyderegulationtexas.com/energy-deregulation-texas/.
78. Ibid.
79. Retrieved from http://fuelfix.com/blog/2013/12/26/deregulation-has-brought-22-billion-more-in-electric-bills-for-.
80. See Roger H. Coupal and David Holland, "Economic Impact of Electric Power Industry Deregulation on the State of Washington: A General Equilibrium Analysis," *Journal of Agriculture and Resource Economics*, 27, 1 (2002): 244–260.
81. Ohio Senate Bill 3, 1999, retrieved from http://www.legislature.state.oh.us/BillText123/123_SB_3_10_N.htm.
82. Ohio Senate Bill 221, retrieved from http://www.legislature.state.oh.us/bills.cfm?ID=127_SB_221.
83. See generally, Chapter 4901:1–38.
84. Ibid.
85. Ibid.
86. Ibid.
87. Ibid.
88. Ibid.
89. Ibid.
90. Casey Junkins, "Ormet Bill Introduced," *Wheeling News-Register*, February 6, 2014.
91. US Energy Information Administration, 2011 Annual Energy Report, p. 37.
92. Ibid.
93. Ibid.
94. "Compressed Natural Gas Infrastructure: Final Report," *America's Natural Gas Alliance*, (Washington, DC) 2012.
95. Ibid., p. 10.

96. Ibid.
97. Ibid.
98. Ibid., p. 12.
99. Ibid., p. 14.
100. Ibid.
101. Ibid.
102. Ibid.
103. Ibid., p. 48.

7 LAND USE AND ENERGY LED ECONOMIC DEVELOPMENT

1. Robert Frellich and Neil Popowitz, "The Umbrella of Sustainability: Smart Growth, New Urbanism, Renewable Energy and Green Development in the 21st Century," *Urban Lawyer*, 42, 1 (Winter 2010): 1–39.
2. Ibid.
3. Mark Muro and Robert Puentes, *Investing in a Better Future: A Review of the Fiscal and Competitive Advantages of Smarter Growth Development Patters* (Washington, DC: Brookings Institution, March 2004).
4. Ibid.
5. Ibid.
6. *15 Economic Facts about Millennials* (Washington, DC: The White House, The Council of Economic Advisers, October 2014).
7. Ibid.
8. Fred Dews, "11 Facts about the Millennial Generation," Brookings Institute, June 2, 2014, retrieved from http://www.brookings.edu/blogs/brookings-now/posts/2014/06/11-facts-about-the-millennial-generation.
9. *15 Economic Facts about Millennials*, The White House.
10. Vivian Giang, "Here's How Millennials Really View Loyalty in the Workplace," *Business Insider*, September 17, 2012, retrieved from http://www.businessinsider.com/how-millennials-really-view-loyalty-2012-9#ixzz3ITmGwUli.
11. Tony Dutzik, Jeff Inglis, and Phineas Baxandall, *Millennials in Motion Changing Travel Habits of Young Americans and the Implications for Public Policy* (Boston: US PIRG Education Fund & Frontier Group, 2014).
12. Ibid.
13. Ibid.
14. Ibid.
15. Retrieved from http://www.nielsen.com/content/corporate/us/en/insights/news/2014/millennials-prefer-cities-to-suburbs-subways-to-driveways.html.
16. Katherine Peralta, "Homeownership for Millennials Declines to New Lows New Data Show Slight Drop in Homeownership for Americans under 35," *U.S. News & World Report*, April 30, 2014.

17. Bana Jobe, "6 Ways Millennials Will Change the Energy Industry from the Inside Out," September 4, 2014, http://thewire.oncor.com/News/Pages/6-Ways-Millennials-Will-Change-the-Energy-Industry-from-the-Inside-Out.aspx.
18. Ibid.
19. Ibid.
20. Ibid.
21. Roger L. Kemp, "The American Dream Moves Downtown," retrieved from http://www.wfs.org/content/american-dream-moves-downtown.
22. Joe Gose, "Millennials Going to Kansas City, to Live and Work," *New York Times*, August 19, 2014.
23. Ibid.
24. Ibid.
25. Ibid.
26. Ibid.
27. Andres Duany, Elizabeth Plater–Zyberk, and Jeff Speck, *Suburban Nation: The Rise of Sprawl and the Decline of the American Dream* (Union Square West, NY: North Point Press, 2000), pp. 5–7.
28. See generally, E. R. Alexander, "A Transaction-Cost Theory of Land Use Planning and Development Control: Towards the Institutional Analysis of Public Planning," *The Town Planning Review*, 72, 1 (January 2001): 45–75.
29. Peter Katz, "Sarasota's Smart Growth Dividend," Planning. American Planning Association, December 2010.
30. Muro and Puentes, *Investing in a Better Future*.
31. Federal Highway Administration, "Case Study: Envision Utah," http://www.fhwa.dot.gov/planning/toolbox/utah_overview.htm. Accessed May 21, 2012.
32. Cameron Speir and Kurt Stephenson, "Does Sprawl Cost Us All? Isolating the Effects of Housing Patterns on Public Water and Sewer Costs," *Journal of the American Planning Association*, 68, 1 (2002): 56–70.
33. Masayuki Morikawa, "Population Density and Efficiency in Energy Consumption: An Empirical Analysis of Service Establishments," *Energy Economics*, 34, 5 (September 2012): 1617–1622.
34. Peter Newman and Jeffrey Kenworthy, "Urban Design to Reduce Automobile Dependence," *Opolis: An International Journal of Suburban and Metropolitan Studies*, 2, 1 (2006): 35–52, Article 3.
35. Retrieved from http://people.hofstra.edu/geotrans/eng/ch8en/conc8en/urbanenergy.html.
36. Reid Ewing, Rolf Pendall, and Don Chen, *Measuring Sprawl and Its Impact* (Washington, DC: Rutgers University, Smart Growth America, 2002).
37. Retrieved from http://www.asce.org/Press-Releases/2011/New-Report-Shows-Failing-to-Invest-in-Transportation-Will-Cause-Job-Losses,-Shrink-Household-Incomes/.
38. Frellich and Popowitz, "The Umbrella of Sustainability," pp. 1–39.
39. Retrieved from http://www.asce.org/Press-Releases/2011/New-Report-Shows-Failing-to-Invest-in-Transportation-Will-Cause-Job-Losses,-Shrink-Household-Incomes/.

40. See generally, "King County, Washington Definition of Development Right," retrieved from http://www.kingcounty.gov/environment/stewardship/sustainable-building/transfer-development-rights/definitions.aspx.
41. Ibid.
42. Ibid.
43. Ibid.
44. Ibid.
45. Brad Berton, "Giving as Good as They Get: Zero-Energy Districts," Urban Land, February 21, 2013, retrieved from http://urbanland.uli.org/sustainability/giving-as-good-as-they-get-zero-energy-districts/.
46. Ibid.
47. Ibid.
48. Ibid.
49. Ibid.
50. Ibid.
51. Retrieved from http://www.dsireusa.org/incentives/incentive.cfm?Incentive_Code=NH42F.
52. Ibid.
53. Ibid.
54. US Bureau of Census; and Oregon Office of Economic Analysis.
55. Retrieved from http://www.oregon.gov/LCD/Pages/lcdc.aspx.
56. Ibid.
57. Retrieved from http://www.oregonmetro.gov/urban-growth-boundary.
58. Ibid.
59. Ibid.
60. Ibid.
61. Peter Calthorpe, *The Next American Metropolis: Ecology, Community, and the American Dream* (New York: Princeton Architectural Press, 1993); Peter Katz, *The New Urbanism: Toward an Architecture of Community* (New York: McGraw-Hill, 1994).
62. Retrieved from http://www.nhhsrail.com/pdfs/TODcasestudydraft_100311.pdf.
63. Ibid.
64. Ibid.
65. Retrieved from http://www.thelandbank.org/whatwedo.asp.
66. Retrieved from http://www.thelandbank.org/history.asp.
67. Ibid.
68. Retrieved from http://www.smartgrowthamerica.org/documents/sga_statepolicy_toolkit.pdf.
69. Ibid.
70. Ibid.
71. Retrieved from http://www.epa.gov/swerosps/bf/laws/2869sum.htm.
72. Ibid.
73. Ibid.
74. Retrieved from http://www.epa.gov/swerosps/bf/job.htm.
75. Retrieved from http://www.epa.gov/swerosps/bf/grant_info/rlf/rlf_factsheet.pdf.

76. See http://www.epa.gov/swerosps/bf/index.html.
77. Retrieved from http://www.epa.gov/swerosps/bf/about.htm.
78. See generally, http://www.dep.state.fl.us/waste/categories/brownfields/pages/act.htm.
79. Retrieved from http://www.dep.state.fl.us/waste/categories/vctc/default.htm.
80. Ibid.
81. Retrieved from http://www.dep.state.fl.us/waste/quick_topics/publications/wc/brownfields/Incentives/BrownfieldBonus.pdf.
82. Retrieved from http://www.dep.state.fl.us/waste/quick_topics/publications/wc/brownfields/AnnualReport/2013/2013BrowardReport.pdf.
83. Ibid.
84. See Peter V. Schaeffer and Cecily Ahern Millerick, "The Impact of Historic Designation on Property Values: An Empirical Study," *Economic Development Quarterly*, 5 (November 1991): 301–312.
85. Retrieved from http://www.preservationnation.org/information-center/sustainable-communities/sustainability/green-lab/valuing-building-reuse.html#.ULKA2qPpV8E.
86. Ibid.
87. See John I. Gilderbloom, Matthew J. Hanka, and Joshua D. Ambrosius, "Historic Preservation's Impact on Job Creation, Property Values and Environmental Stability," *Journal of Urbanism: International Research on Placemaking and Urban Sustainability*, 2, 2 (July 23, 2009): 83–101.
88. Ibid.
89. San Mateo City Code, Ch. 27.66, et seq, available at http://www.cityofsanmateo.org/index.asp?NID=808.
90. Retrieved from http://www.npa.gov/tps/tax-incentives.htm.
91. Ibid.
92. Retrieved from http://www.hpo.ncdcr.gov/tchome.htm.
93. Ibid.
94. Ibid.
95. Ibid.

8 Building an Urban Energy Center

1. Bob Herman, "Houston's Energy Company Boom Fuels Hospital Expansion," Modern Health Care, June 28, 2014, retrieved from http://www.modernhealthcare.com/article/20140628/NEWS/306279956.
2. Ibid.
3. Kurt Badenhausen, "While Rest of U.S. Economy Plods, Houston Gets Hot," *Forbes Magazine*, July 6, 2012.
4. *Global Metro Monitoring* (Washington, DC: Brookings Institution, 2011–2012).
5. *The Economy at a Glance Houston* (Houston, TX: Greater Houston Partnership, September 2014).

6. Ibid.
7. Ibid.
8. Ibid.
9. Ibid.
10. Ibid.
11. Ibid.
12. *Houston Region Economic Development Profile* (Houston, TX: Greater Houston Partnership, 2013).
13. *The Economy at a Glance Houston.*
14. Ibid.
15. US Census, Quick Facts, Houston, retrieved from http://quickfacts.census.gov/qfd/states/48/4835000.html.
16. *Houston Region Economic Development Profile.*
17. Ibid.
18. *The Economy at a Glance Houston.*
19. US Census, Quick Facts, Houston, retrieved from http://quickfacts.census.gov/qfd/states/48/4835000.html.
20. Ibid.
21. Ibid.
22. Ibid.
23. Joel Kotkin, "Houston Rising—Why the Next Great American Cities Aren't What You Think," *The Daily Beast*, April 8, 2013, retrieved from http://www.thedailybeast.com/articles/2013/04/08/houston-rising-why-the-next-great-american-cities-aren-t-what-you-think.html.
24. Ibid.
25. *Houston Region Economic Development Profile.*
26. Ibid.
27. Ibid.
28. Ibid.
29. Ibid.
30. Ibid.
31. Ibid.
32. Ibid.
33. Ibid.
34. Ibid.
35. Igor Vojnovic, "Governance in Houston: Growth Theories and Urban Pressures," *Journal of Urban Affairs*, 25, 5 (2003): 589–624, 595.
36. Ibid.
37. Ibid.
38. Retrieved from http://geology.com/articles/history-of-energy-use/.
39. Ibid.
40. Ibid.
41. Robert Wooster and Christine Moor Sanders, "Spindletop Oilfield," *Handbook of Texas Online*, http://www.tshaonline.org/handbook/online/articles/dos03, accessed October 12, 2014. Uploaded on June 15, 2010. Published by the Texas State Historical Association.

42. Ibid.
43. Ibid.
44. Ibid.
45. Ibid.
46. Ibid.
47. Ibid.
48. Ibid.
49. Vojnovic, "Governance in Houston," p. 603.
50. Ibid.
51. Ibid.
52. Ibid.
53. Ibid., p. 606.
54. Ibid.
55. Jean-Paul Rodrigue, *The Geography of Transport Systems*, 3rd ed. (New York: Routledge, 2013).
56. Ibid.
57. Ibid., p. 609.
58. Ibid.
59. Ibid., p. 607.
60. Ibid.
61. Ibid., p. 604.
62. Ibid., p. 609.
63. "Houston," *Fortune*, July 14, 1980, p. 49.
64. *Zoning Guide* (Washington, DC: US Department of Commerce, rev. ed. 1926), §1–§2.
65. Andres Duany, Elizabeth Plater–Zyberk, and Jeff Speck, *Suburban Nation: The Rise of Sprawl and the Decline of the American Dream* (Union Square West, NY: North Point Press, 2000), pp. 5–7.
66. See Chapter 42, Houston City Code (2013) retrieved from http://www.houstontx.gov/planning/DevelopRegs/docs_pdfs/chapter_42_redline_03_18_2013.pdf.
67. Ibid.
68. Ibid.
69. Ibid.
70. Ibid., pp. 38–39.
71. Jeffrey M. Lacker, *Manufacturing in the New Southern Economy* (Roanoke, VA: Federal Reserve Bank of Richmond, Southern Growth's 2011 Chairman's Conference, June 13, 2011).
72. Neil Shah and Ben Casselman, "'Right-to-Work' Economics: States That Bar Mandatory Union Dues Tend toward More Jobs but Lower Wages," *Wall Street Journal*, December 14, 2012, retrieved from http://www.wsj.com/articles/SB10001424127887324296604578179603136860138.
73. *Oklahoma Council of Public Affairs Eliminating the State Income Tax in Oklahoma: An Economic Assessment*, Oklahoma Council of Public Affairs, Arthur Laffer, November 2011.

74. *An Overview of Taxes in Texas,* Texas Taxpayers and Research Association, January 2011.
75. *Houston Region Economic Development Profile.*
76. *An Overview of Taxes in Texas.*
77. Ibid.
78. *Houston Region Economic Development Profile.*
79. Ibid.
80. *An Overview of Taxes in Texas.*
81. Retrieved from http://www.houstontx.gov/ecodev/tirz.html.
82. Ibid.
83. Ibid.
84. Ibid.
85. Ibid.
86. Ibid.
87. Retrieved from http://www.uptown-houston.com/about/page/uptown-tirz-uda.
88. Ibid.
89. Ibid.
90. Ibid.
91. Retrieved from http://www.houstontx.gov/ecodev/abatements.html.
92. Ibid.
93. Ibid.
94. Ibid.
95. Ibid.
96. Ibid.
97. Retrieved from http://www.bizjournals.com/houston/blog/breaking-ground/2013/08/houston-city-council-passes-downtown.html.
98. Retrieved from http://www.portlandonline.com/bps/index.cfm?c=30357.
99. See generally, Scott Houston, "Municipal Annexation in Texas, 'Is It Really That Complicated," *Texas Municipal League,* March 2012.
100. Ibid.
101. Ibid.
102. Ibid.
103. Ibid.
104. Ibid.
105. Ibid.
106. US Census Bureau, 2012 Reports.
107. Retrieved from http://www.houston.org/business/infrastructure.html.
108. Ibid.
109. Ibid.
110. Retrieved from http://www.portofhouston.com/about-us/economic-impact/.
111. Ibid.
112. Ibid.
113. Retrieved from http://www.houston.org/business/infrastructure.html.
114. Ibid.
115. Ibid.

116. Steven Craig, *Houston Airport System Economic Impact Study, Final Report* (Jenkintown, PA: GRA, Incorporated, June 30, 2011).
117. Ibid.
118. Ibid.
119. Ibid.
120. Retrieved from http://www.houston.org/business/infrastructure.html.
121. Ibid.
122. Ibid.
123. Ibid.
124. Greater Houston Partnership, Workforce Fact Sheet, retrieved from https://www.houston.org/economy/index.html#Workforce.
125. Ibid.
126. Ibid.
127. Ibid.
128. Ibid.
129. Ibid.
130. *The STEM Workforce Challenge*, US Department of Labor, 2007.
131. Bureau of Labor Statistics, 2012.
132. Greater Houston Partnership, Workforce Fact Sheet, retrieved from https://www.houston.org/economy/index.html#Workforce.
133. Ibid.
134. Ibid.
135. Ibid.
136. Retrieved from http://www.demographia.com/db-uscityr1900.htm.
137. David R. Eichenthal, *Why Mid-Size Cities Grow and What It Means for Chattanooga* (Washington, DC: Brookings Institution Center on Metropolitan Policy, September 2005).

9 Building a Green Economy

1. Union of Concerned Scientist, Benefits of Renewable Energy, 2014, retrieved from http://www.ucsusa.org/clean_energy/our-energy-choices/renewable-energy/public-benefits-of-renewable.html#.VEacRLl0zcs.
2. William M. Bowen, Sunjoo Park, and Joel A. Elvery, "Empirical Evidence of the Influence of Renewable Energy Portfolio Standards on the Green Economies of States," *Economic Development Quarterly*, 27, 4 (2013): 338–351, 339.
3. Ibid.
4. Ibid.
5. Ibid.
6. Ibid.
7. Ibid.
8. Ibid.

9. Mark Muro, Jonathan Rothwell, and Devashree Saha, *Sizing the Clean Economy: A National and Regional Green Jobs Assessment* (Washington, DC: Brookings Institution, July 13, 2011).
10. Ibid.
11. Ibid.
12. Ibid.
13. Eric Lantz, Frank Oteri, Suzanne Tegen, and Elizabeth Doris, "State Clean Energy Policies Analysis: State Policy and the Pursuit of Renewable Energy Manufacturing," *National Renewable Energy Laboratory Technical Report NREL/TP 6A2–46672*, February 2010.
14. Ibid.
15. Ibid.
16. Ibid.
17. Ibid.
18. Michael C. Slattery, Becky L. Johnson, Jeffrey A. Swofford, and Martin J. Pasqualletti, "The Predominance of Economic Development in the Support of Large-Scale Wind Farms in the U.S. Great Plains," *Renewable and Sustainable Energy Review*, 16 (2012): 3690–3701, 3691.
19. Michael C. Slattery, Eric Lantz, and Becky L. Johnson, "State and Local Economic Impacts from Wind Energy Projects: Texas Case Study," *Energy Policy*, 39 (2011): 7930–7940, 7932.
20. Ibid.
21. Ibid.
22. Ibid., p. 7934.
23. Muro, Rothwell, and Saha, *Sizing the Clean Economy.*
24. Ibid.
25. Juita-Elana (Wie) Yusof and Katherine A. Neill, "State Energy-Based Economic Development Policies and Examples," *Economic Development Quarterly*, 27, 3 (2013): 240–249, 241.
26. See Database for State Incentives Renewable and Efficiency, retrieved from http://www.dsireusa.org/glossary/.
27. *Guide to Federal Financing for Energy Efficiency and Clean Energy Deployment* (Washington, DC: Department of Energy, September 2014).
28. Ibid.
29. Ibid.
30. Ibid.
31. Ibid.
32. Ibid.
33. Ibid.
34. Ibid.
35. Ibid.
36. Ibid.
37. Ibid.
38. Ibid.
39. Ibid.

40. Ibid.
41. Ibid.
42. Ibid.
43. Ibid.
44. Ibid.
45. Ibid.
46. Ibid.
47. Ibid.
48. Ibid.
49. Ibid.
50. Ibid.
51. Ibid.
52. Ibid.
53. Ibid.
54. Ibid.
55. Ibid.
56. Ibid.
57. Ibid.
58. Ibid.
59. Ibid.
60. Ibid.
61. Ibid.
62. Ibid.
63. Ibid.
64. Ibid.
65. Ibid.
66. Ibid.
67. See Database for State Incentives Renewable and Efficiency, retrieved from http://www.dsireusa.org/glossary/.
68. Ibid.
69. Ibid.
70. Ibid.
71. Ibid.
72. Ibid.
73. Ibid.
74. Ibid.
75. Ibid.
76. Ibid.
77. Ibid.
78. Ibid.
79. Ibid.
80. Ibid.
81. Ibid.
82. Ibid.
83. US Census Bureau, Oregon Quick Facts, 2013, retrieved from http://quickfacts.census.gov/qfd/states/41000.html.

84. Ibid.
85. Ibid.
86. City Data, 2013, retrieved from http://www.city-data.com/city/Portland-Oregon.html.
87. Ibid.
88. Ibid.
89. Pew Center, Oregon Clean Energy Economy Report, 2008, retrieved from http://www.oregonbusinessplan.org/LinkClick.aspx?fileticket=tCHHCEQfudU%3d&tabid=211-Pew Center-Oregon Clean Energy Economy.
90. Ibid.
91. Ibid.
92. Ibid.
93. Clean Edge Clean Energy Oregon Report, 2012, retrieved from https://cleanedge.com/reports/Oregons-Clean-Energy-Economy.
94. Ibid.
95. Ibid.
96. Ibid.
97. American Council on Renewable Energy, Oregon Fact Sheet, 2014, retrieved from http://www.acore.org/files/pdfs/states/Oregon.pdf.
98. Ibid.
99. Ibid.
100. Ibid.
101. Ibid.
102. Ibid.
103. Ibid.
104. Ibid.
105. Ibid.
106. Retrieved from http://dsireusa.org/incentives/incentive.cfm?Incentive_Code=OR153F&re=0&ee=0.
107. Ibid.
108. Ibid.
109. Ibid.
110. American Council on Renewable Energy, Oregon Fact Sheet, 2014, retrieved from http://www.acore.org/files/pdfs/states/Oregon.pdf.
111. Ibid.
112. Ibid.
113. Ibid.
114. Retrieved from http://dsireusa.org/incentives/incentive.cfm?Incentive_Code=OR153F&re=0&ee=0.
115. Ibid.
116. Ibid.
117. Ibid.
118. Ibid.
119. Ibid.
120. Ibid.
121. Ibid.

122. Ibid.
123. Ibid.
124. Ibid.
125. Ibid.
126. Ibid.
127. Ibid.
128. Ibid.
129. Ibid.
130. Ibid.
131. Ibid.
132. Ibid.
133. Ibid.
134. Ibid.
135. Ibid.
136. Ibid.
137. Ibid.
138. Ibid.
139. Ibid.
140. Ibid.
141. Ibid.
142. Ore. Rev. Stat. Sec. 197.005–197.860 (1991).
143. Ibid. at Sec. 197.010, 197.175.
144. Ibid. at Sec. 197.030.
145. Ibid.
146. Ibid. at Sec. 197.299.
147. Ibid.
148. Stuart Meck, Ed., *American Planning Association, Growing Smart Legislation Guidebook: Model Statutes for Planning and the Management of Change, Note 6C: A Note on Existing Regional Plans* (Chicago, IL: American Planning Association, 2002).

10 Shale Energy Revolution

1. US Census Bureau, American Fact Finder, 2008.
2. Allen S. Chambers, "Capital Center and South Central West Virginia," in *Buildings of West Virginia* (New York: Oxford University Press, 2004), pp. 114–115 ISBN 0–19–516548–9.
3. http://www.abandonedonline.net/2013/06/11/the-ghost-town-of-thurmond-west-virginia/
4. Ibid.
5. "Strengthening the Rural Economy—The Current State of Rural America," White House Council of Economic Advisors, retrieved from www.whitehouse.gov/administration/eop/cea/factsheets-reports/strengthening-the-rural-economy.

6. Ibid.
7. Ibid.
8. Ibid.
9. Ibid.
10. Ibid.
11. Kim Severson, "Digital Age Is Low to Arrive in Rural America," *The New York Times*, February 11, 2011.
12. Retrieved from http://www.willistondevelopment.com/.
13. See John Nolan, "Land Use and Sustainability: Is There Hope for Rural America," *Plenary Address, Preventing the Ghost Town: What Rural Communities Need to Do to Survive in the Modern Economy*, February 21, 2014.
14. Ibid.
15. Drew Mackenzie, "North Dakota Oil Boom Fuels Big Growth in Millionaires," *Newsmax*, January 17, 2014.
16. Ibid.
17. Ibid.
18. Brian Louis, "Shale Boom Lures Developer for $500 Million North Dakota Project," *Bloomberg News*, May 20, 2014.
19. *Eagle Ford Shale Economic Impact for Counties with Active Drilling*, Center for Community and Business Research, University of Texas at San Antonio, Institute for Economic Development, San Antonio, Texas, October 2012.
20. Ibid.
21. Ibid.
22. Ibid.
23. Ibid.
24. Ibid.
25. Timothy W. Kelsey, Martin Shields, James R. Ladlee and Melissa Ward, "Economic Impact of Marcellus Shale in Pennsylvania: Employment and Income in 2009," Marcellus Shale Education & Training Center, August 2011.
26. Joe Massaro, "Marcellus Shale a Source of Major New Revenues for Local Townships," Energy Indepth, June 6, 2014.
27. Ibid.
28. Kelsey, Shields, Ladlee, and Ward, "Economic Impact of Marcellus Shale in Pennsylvania."
29. Dan Black, Terra McKinnish, and Seth Sanders, "The Economic Impact of the Coal Boom and Bust," *The Economic Journal*, 115 (April 2005): 449–476.
30. Ibid.
31. Sean O'Leary and Ted Boettner, "Boom and Busts: The Impact of West Virginia's Energy Economy," West Virginia Center on Budget & Policy, July 2011.
32. Ibid.
33. "West Virginia Profile," *Regional Profile*, Federal Reserve Bank of Richmond, August 2013, retrieved from http://www.richmondfed.org/research/regional_economy/reports/regional_profiles/pdf/wv_regional_profile.pdf.
34. Ibid.

35. O'Leary and Boettner, "Boom and Busts."
36. Retrieved from http://stateimpact.npr.org/pennsylvania/tag/bradford-county/.
37. See retrieved from http://www.willistondevelopment.com/.
38. See generally the City of Indianapolis-Marion County Comprehensive Plan, retrieved from http://www.indy.gov/eGov/City/DMD/Planning/Services/Land/Pages/comp_plan.aspx.
39. Ibid.
40. See http://www.eda.gov/Research/ClusterBased.xml.
41. Ibid.
42. Ibid.
43. Michael E. Porter, "Location, Competition and Economic Development: Local Clusters in a Global Economy," *Economic Development Quarterly*, 14, 1 (2000):15–34.
44. Ibid.
45. *2013 Annual Act 13—Unconventional Gas Well Impact Fee Report to the Pennsylvania Legislature and the Governor's Budget Office*, Pennsylvania Utility Commission, January 1–December 31, 2013.
46. Pennsylvania Act 13 of 2012.
47. Ibid.
48. Ibid.
49. Ibid.
50. Ibid.
51. Ibid.
52. Ibid.
53. Ibid.
54. See Rose Naccarato, "Tax Increment Financing, Opportunities and Concerns," *Staff Research Brief, Tennessee Agency on Intergovernmental Relations*, March 2007, retrieved from http://www.tn.gov/tacir/PDF_FILES/Taxes/Tax%20Increment%20Financing.pdf.
55. *How TIF Works: Basic Mechanics*, Minnesota House of Representatives, House Research, 2013, retrieved from http://www.house.leg.state.mn.us/hrd/issinfo/tifmech.htm.
56. 70C AM. JUR. 2d Special or Local Assessments § 22, 2008.
57. Ohio Water Development Authority 2013 Annual Report, retrieved from http://www.owda.org/owda-doc/AnnualReport2013.pdf.
58. Ibid.
59. Malcom Hall, "Carrollton Water Plant Ready to Go with the Flow," *Canton Repository*, March 26, 2014.
60. Retrieved from http://www.ferc.gov/industries/gas/indus-act/pipelines.asp.
61. Ibid.
62. Retrieved from http://www.ferc.gov/industries/gas/indus-act/pipelines/pending-projects.asp.
63. Retrieved from http://www.ferc.gov/industries/gas/indus-act/pipelines.asp.
64. Retrieved from http://www.ferc.gov/for-citizens/citizen-guides/citz-guide-gas.pdf.

65. "American Industry and Fracking, from Sunset to New Dawn," *The Economist*, New York, November 16, 2013.
66. Ibid.
67. Ibid.
68. Ibid.
69. Retrieved from http://aeped.com/business-development-services/qsp/aep-quality-sites/.
70. Ibid.
71. Paul Glader, "A $70 Million Rec Center Brings Indoor Surfing, Golf to North Dakota," *Business Week*, June 18, 2014.
72. Ibid.
73. Ibid.
74. Retrieved from http://www.dleg.state.mi.us/mpsc/comm/broadband/broadband/broadband.htm.
75. Ivan T. Kandilov and Mitch Renkow, "Infrastructure Investment and Rural, Economic Development: An Evaluation of USDA's Broadband Loan Program," *Growth & Change*, 41, 2 (June 2010): 165–191.
76. Ibid.
77. Ibid.
78. Ibid.
79. Retrieved from http://www.portal.state.pa.us/portal/server.pt/community/act_13/20789/natural_gas_vehicle_program/1157504.
80. Ibid.
81. Ibid.
82. Ibid.
83. Ibid.
84. Ibid.

Index

9781137471697